ENVIRONMENTAL ARCHAEOLOGY

PRINCIPLES AND METHODS

John Evans and Terry O'Connor

SUTTON PUBLISHING

First published in 1999 by
Sutton Publishing Limited · Phoenix Mill
Thrupp · Stroud · Gloucestershire · GL5 2BU

Reprinted in 2001

Copyright © John Evans and Terry O'Connor, 1999

All rights reserved. No part of this publication may be reproduced, stored in a retrieval system, or transmitted, in any form, or by any means, electronic, mechanical, photocopying, recording or otherwise, without the prior permission of the publisher and copyright holders.

The authors have asserted the moral right to be identified as the authors of this work.

British Library Cataloguing in Publication Data
A catalogue record for this book is available from the British Library

ISBN 0 7509 1779 2

Typeset in 10/13pt Sabon.
Typesetting and origination by
Sutton Publishing Limited.
Printed in Great Britain by
J.H. Haynes & Co. Ltd, Sparkford.

Contents

List of Figures — vii
List of Tables — x
Preface — xi
Acknowledgements — xiii

1 INTRODUCTION — 1

Box 1.1 *Chronology of developments in archaeology and environmental studies* 2

———— PART I: LIFE ————

2 HUMAN ENVIRONMENTS — 10
Lithosphere 11   Atmosphere 12   Topography 13   Biosphere 13   People 15

3 ECOLOGY AND ENVIRONMENT — 17
The Biosphere 17   Species 17   Species' ranges 17   Within-range variation 18
Across-species variation 21   Mobility and aggregation 21   Niches 23   Abundance 23
Multiple-species communities 26   Classification of the biosphere 28   **Soil Processes** 29
Soil formation 29   Soil processes and soil horizons 32   Soil profiles and soil types 32
Environmental groupings 33   Other groupings 35   Complications 36   Summary of
groupings 37   **Sedimentation** 37   General properties 37   Characteristics of
sedimentation 38   Classification of sedimentation processes 39   The onset, continuation
and cessation of sedimentation 41   Sediment classification and terminology 41
Sediment units 41   Spatial distribution and sequences 43   A word of warning 44

Box 3.1 *r- and K-strategists* 20
Box 3.2 *Temporo-spatial variations in seabirds and waterfowl* 24
Box 3.3 *What do we mean by 'niche'?* 27
Box 3.4 *Soil horizons and soil types* 30

4 ECOSYSTEMS — 45
Definition and Characteristics of Ecosystems 45   Space 46   Time 46   Gradients 47
Succession 51   Humans 53   A Classification of Ecosystems 54   Changing Fundamentals 54

Box 4.1 *Ecosystem changes in the Holocene* 55

CONTENTS

———————— PART II: FROM DEATH TO BURIAL ————————

5 A CRANNOG AND ITS ENVIRONMENT — 60
Different Categories of Materials 60   Different Scales of Space 60   The site 62
The locality 64   The region 64   The wider environment 64   **Relevance of the Spatial Integrity of the Data** 65   The Special Nature of the Site 65   The Site Catchment 66

6 DEATH ASSEMBLAGE FORMATION — 67
Death of Organisms 67   Attritional death 68   Death-traps or locationally biased death 68   The effects of predators and scavengers 69   Catastrophic death 69   Hominid- and human-originated death 69   **Modelling Death Assemblages** 72   Community Death 72   Ecosystem Abandonment 76   Ecosystem Death by Burial 76

Box 6.1 Modelling pollen assemblages 70
Box 6.2 Modelling bone death assemblages 74

7 DEPOSITION AND POST-DEPOSITION — 78
The Nature of the Archaeology 78   Physico-Chemical Conditions 81   Environmental Landblocks 82   **Movement and Change** 82   Lateral movement 82   Minimal or no lateral movement 86   **Post-Deposition Transformation of Land and Archaeology** 89   Zones of Destruction and Preservation 91   **Representativeness of Preservation** 92

Box 7.1 Physico-chemical conditions of preservation 80

———————— PART III: PROCEDURES AND METHODS ————————

8 RESEARCH DESIGNS AND SAMPLING STRATEGIES — 94
The Global View 94   Spatial Areas of Research 94   How not to define areas of research 95   Defining areas of research 96   **Intra-Areal Sampling** 101   The physical environment 101   Questions 102   Why? 103   Small-Scale Strategies 103   Methods 104

Box 8.1 South Nesting, Shetland: scales of sampling 98

9 METHODS OF EXAMINING ARCHAEOLOGICAL DISTRIBUTIONS — 106
Recording Systems 106   Modern Maps 106   Aerial Photography 109   Historical Documents and Toponymy 109   Maps 110   Documents 110   Toponymy 111   Ground-Level Survey 112   Above-ground survey 112   Fieldwalking 114   Phosphate and Magnetic Susceptibility Measurement 115   Geophysics 115   Biogeography 116   Conclusions 117

10 ANALYSIS AND USE OF SOILS AND SEDIMENTS — 118
Mapping 118   Field Description 120   Bulk Laboratory Analysis 121   Micromorphology 122   **Soils and Sediments in Archaeology** 127   Taphonomy 127   Soils and past land-use 128   Soil and sediment histories 129

Box 10.1 Distinctions between soils and sediments 119
Box 10.2 The prediction of site distributions 126

11 BIOLOGICAL INDICATORS — 132
General Issues 132   Plant Microfossils 134   Pollen 134   Phytoliths 135   Diatoms 136   Cereal bran 136   **Plant Macrofossils** 137   **Wood** 139   **Invertebrates** 140   Beetles 140   Molluscs 141   Other invertebrates 144   **Vertebrates** 145

Box 11.1 Stellmoor, Germany: integrating freshwater molluscs and ostracods 142

12 CONTEXTS — 148
Molecules, Isotopes and Increments 148   Archaeological Features and Sites 149   Buried soils 149   Ditches, pits and wells 151   Occupation layers 151   Tells 152   Hillforts 153   **Big Areas** 154   Uplands 154   River valleys 160   Coasts and estuaries 165

PART IV: INTERPRETATION

---

13 DATA ANALYSIS — 171
Grouping Data Internally 172   Nearest neighbour analysis 172   Spatial associations 172   Diversity and equitability 172   Methods of assemblage comparison 175   **Methods which use Individual Species' Ecology** 177   Ecological groupings 177   Phytosociology 178   Mutual climatic range method (MCRM) 179

14 POSSIBILISTIC STUDIES — 181
Human-Life Studies 182   Ethnoarchaeology 185   Experiment 185   Ethnoarchaeology, Experiment and Taphonomy 188   Site Exploitation Territorial Analysis 189   Thiessen Polygons 194   Mathematical Modelling and Computer Simulation 194

Box 14.1 Howmore, South Uist: ethnoarchaeology case study 1 186
Box 14.2 The Dassenetch: ethnoarchaeology case study 2 187
Box 14.3 SETA in the Upper Palaeolithic of northern Spain 190
Box 14.4 A GIS case study 192

15 CASE STUDIES — 196
Northern England: The Roman Impact 196   The Earlier Neolithic of the English Chalk 201   Procedural stages of molluscan interpretation 202   The environmental sequence 205   What else should we be doing? 207   The absence of Neolithic erosion 208   Later erosion 209   Conclusions 209   **Soils and Human Land-Use** 209   Terrace agriculture in semi-arid mountainous localities 209   Soil/human relationships on a Mediterranean karst island 210   The macro-regional view 211

16 THE HUMAN NICHE: A BASIC UNIT OF STUDY FOR ARCHAEOLOGY — 215

References — 219
Index — 237

For Darwin and Guinness

LIST OF FIGURES

Frontispiece	The differentiation of Nature and Culture	xiv
1.1	Catastrophic deposition at Garth's Voe, Shetland	1
2.1	Generalised diagram of sediments from continental interior to ocean floor	12
2.2	Relationship of some major plant groupings to climate at the present day	14
2.3	Categories of environment in relation to people	16
3.1	Ranges of the same species in two different climatic conditions	19
3.2	Distribution of shell shape in various gastropod (snail) groups	22
3.3	Pyramids of energy of plants and animals in immature summer deciduous woodland	26
3.4	Translocation of materials and chemicals in soil profiles	33
3.5	Tracks of soil formation in the later Holocene in a temperate climate region	34
3.6	Argillic brownearth profile on Carboniferous Limestone, Malham, Yorkshire	35
3.7	Complex soil profile, typical of British uplands around the limits of cultivation	36
3.8	The general relationships of soil and different categories of sediment	39
3.9	Relationship between conventional soils and sediments and the archaeological and biological remains within them	40
3.10	Oyster shell outer surface, providing a substratum for boring worms, marine snails and bivalves	40
3.11	Section of travertine from Marsworth, Herts	41
4.1	Four levels of time, to show the potential for change in resilient and fragile ecosystems	47
4.2	Variation of sediments and land at the sub-continental scale in relation to an ice sheet	48
4.3	West-east transect at the eastern end of the Mediterranean	49
4.4	Variation of sediments and land at the intra-regional scale in a temperate maritime region	49
4.5	Variation of soil and land in a non-glaciated area	50
4.6	Land variation at the local scale in a temperate climate region at the limits of cultivation; Gordale, near Malham, Yorkshire	51
4.7	Relationship between vegetation and sediments at a lake edge	52
II.1	Stages from ancient life to archaeological publication	57
II.2	Stages in the history of the relationships of a pot	58
5.1	Section through a lake and its surrounds in relation to a crannog	61
5.2	Section through a part of a lake and its deposits showing different habitats	63
6.1	The contrast between animals/plants and humans in regard to their preserved remains and study	67

LIST OF FIGURES

6.2	The decrease in temporo-spatial structure of an animal community from life through death to burial	73
7.1	Different kinds of archaeological signal in relation to past human behaviour	78
7.2	Different categories of human activity in terms of time and space	79
7.3	Different origins of land snails in a ditch	83
7.4	Section of an Iron Age storage pit at Balksbury, Hants	85
7.5	Different stages in the history of relationships of a bone relevant to its place of deposition	86
7.6	Changes from a stable land surface to an unstable land surface with colluviation under arable	87
7.7	Valley section at Avebury, Wilts	88
7.8	The different origins of materials in a soil and sediment	88
7.9	Generalised slope showing decreasing surface artefact diversity from slope to valley bottom	89
7.10	Plan and section of a chalkland valley showing zones of archaeology, land-use and preservation	90
8.1	A river valley is a poor unit of regional research as seen in different distributions of settlement units	96
8.2	Eroded hard limestone pavement, The Burren, Co. Clare, Ireland	97
8.3	Blanket peat accumulations on sandstone, north Co. Mayo, Ireland	97
8.4	A planning aid to sampling strategy for two regions	100
8.5	Sampling strategy for a chalkland area	101
8.6	Distance between data and inference (a) for the same inference, (b) the same data	104
9.1	A strategy for sampling an area within a region	107
9.2	Solid and drift geology and soil maps for the same area	108
9.3	Use of a variety of data in sorting out land chronology	112
9.4	Methods of prospecting and survey	113
9.5	The use of infra-red thermal imaging	113
9.6	Interpretation of stone artefacts found in field-walking	115
9.7	Use of the gradiometer	116
10.1	Soil profile showing some descriptive characters and sampling methods	120
10.2	Soil profile showing categories of sampling	122
10.3	Photomicrograph of a soil thin section from a buried soil	123
10.4	Annotation of photomicrograph in Fig. 10.3	123
10.5	Sequence of pedozones from a tree hollow	125
10.6	The non-equivalence of different stratigraphies from evidence around a lake and in its sediments	129
10.7	Different levels of causes and requirements of soil erosion and deposition	130
10.8	Stages in the formation of fine overbank alluvium	130
11.1	*Vertigo antivertigo* and *V. moulinsiana*, showing adults and juveniles	132
11.2	Taphonomic steps in the preservation of plant macrofossils by charring	138
12.1	Different locations of oxygen isotopes	149
12.2	Bank and ditch sequence from the South Street long barrow, Wilts	150
12.3	Sampling localities in an Iron Age hillfort	154

LIST OF FIGURES

12.4	A quarry hollow at Maiden Castle hillfort	155
12.5	Upland archaeological contexts by zones	156
12.6	A small upland topogenous mire in Cumbria	158
12.7	Correlation of upland agriculture, lake sediments and river terraces	159
12.8	Transverse section of a lowland river valley showing Late-glacial and Holocene contexts	160
12.9	Transverse section of chalkland dry valley showing a basic sequence of colluvium and soils	162
12.10	Transverse section of an upland river valley, with Late-glacial and Holocene depositional contexts	163
12.11	Classification of river valley contexts for the Holocene	163
12.12	Sequence of coastal deposits outside the area of isostatic recovery	165
12.13	Transverse section from upland to coast showing obvious and well-studied loci, and cryptic and under-studied ones	167
12.14	Dalmore, Lewis, from the air	168
IV.1	Relationship between soils/sediments/populations and the proximal and distal environmental factors to which their behaviour relates	169
13.1	Rank order curves and Shannon-Wiener diversity indices for four land mollusc assemblages	173
13.2	Land diversity based on two different calculations of the same area	174
13.3	The use of ordination and similarity indices in grouping assemblages and characterising them by feedback	176
13.4	Principles of the MCRM	179
14.1	The position of possibilistic methods in interpretation	181
15.1	Northern England, with sites mentioned in the text	197
15.2	Two alternative routes and mechanisms in the deposition of arboreal pollen into deposits of the Roman sewer, Church St, York	200
15.3	The headwaters of the River Kennet, central southern England, showing sites mentioned in the text and their main land-use	202
15.4	Buried soil under the Neolithic bank of the Avebury henge	203
15.5	The island of Hvar, Croatia	211

LIST OF TABLES

Page:

Page	
23	Table 3.1. Differences in one species between the edge and middle of its range
38	Table 3.2. Classification of clastic sedimentation processes and the sediments
42	Table 3.3. Sediment classification
61	Table 5.1. Categories of data from a crannog, and the environments and scales of space to which they refer
62	Table 5.2. Different kinds of environmental materials of use at specific scales of reference, all from the site
72	Table 6.1. Summary of types of animal death assemblage formation
83	Table 7.1. Deposition and post-deposition of biological materials, soils and sediments
91	Table 7.2. Zones of archaeology and environment in western Britain on an altitudinal catena from high upland plateau above the tree-line to the coast (Caseldine 1990) (cf. Fig. 4.4)
107	Table 9.1. Various uses of GIS
124	Table 10.1. Pedozones in the infilling of a hypothetical tree-throw pit beneath an archaeological earthwork; 1 is the earliest (cf. Fig. 10.5)
157	Table 12.1. Contexts for Bronze Age archaeology in an upland area, specifically the North York Moors, based on Spratt and Simmons (1976); units arranged in order of descending altitude
206	Table 15.1. Early environments associated with sites in the chalkland study area

PREFACE

This book is about the biophysical environment of past humans and its study. Although there are several books which deal with this, none of them provides a relatively short introduction to the subject which covers theory, methodology and practice as this one does, and none of them explicitly integrates these with archaeology. In addition, our book takes an ecological approach, seeing humans as a part of ecosystems, and encompasses this in their study.

There are several ways in which this subject could be presented at the undergraduate level. We could have started by looking at the complexity of interactions between humans and their environment, and seeing environment in its broadest sense as including the intrinsic human biotic, psychological and social components as well as the extrinsic biophysical ones. We could then have gone on to discuss the problems of separating these two groups of factors as influences in cultural diversity and change. We have chosen not to do this because our subject areas are in the natural sciences: there is, however, certainly room for such a study. Alternatively, we could have used several fundamental areas of the relationship between humans and their environment as the main framework, such as atomic and molecular studies, morphometric and traumatic variations of the human skeleton, the spread of humans into new lands and the changes they wrought in plant and animal populations, but we felt that there would not be enough cohesion in this approach and that it would devolve into a series of case studies. A third possibility, that of presenting biological, soil and sediment techniques, followed by examples of their use in working out past climates, sea-levels and other abstract aspects of the environment we felt would not do justice to the complexity of the subject as an archaeological study, while a fourth, looking explicitly at biophysical environmental change during the period of human history, has been well accomplished by other texts (e.g. Roberts 1989; Bell and Walker 1992).

Instead, what we have chosen to do is present the subject in a way in which it might be conceived by someone doing research, although even here our approach is only one of several which could be used. Much depends on the methodology and interpretation of the researcher. It is our firm belief that experiential iteration of data, impressions and interpretation is the only way forward, just as it has been through human (and, perhaps, earlier hominid) history. We have also avoided a closely site-focused approach because this would have limited the diversity of techniques and contexts which could be discussed.

A short introduction (chapter 1) outlines the history of the subject. There are then four main parts. Part I (chapters 2–4) deals largely with the living aspects of human ecosystems which are preserved as such in the archaeological record, this being a prelude to one part of our interpretational procedure, namely an analogue one. We also introduce the idea that data, however small, embody a complexity of time and space information beyond the physical environment in which they exist. In Part II (chapters 5–7), we look at some of the ways in which these components of ecosystems come to be preserved in the archaeological record, stressing not just the transformations of individual items but of terrain and topography too. Here the approach explicitly combines information from the past itself as well as observations on modern processes. We also stress that all stages in these transformations can be of relevance to past human lives and none is to be seen solely as a means to the interpretation of another. Part III (chapters 8–12) looks at the processes of data-gathering and study of various groups of material data preserved in the archaeological record. This is where the regional approach comes to the fore, the methods being explicitly presented in terms of a regional research strategy. The impossibility of separating data-gathering from interpretation right from the start of the research process is stressed. Indeed, in Part IV (chapters 13–15), the chapters on interpretation could well have been integrated earlier on. They are about different levels of interpretation, chapter 13 on data simplification and grouping, chapter 14 on the use of present-day human situations which elucidate the data, and chapter 15 a series of case studies in which we use both approaches. An end chapter (16) offers a theoretical nexus for the subject as we see it.

This presentation is a progressional one, not just along the research process but also in terms of complexity. Readers new to the subject are advised to read the earlier chapters first; those with some knowledge of Holocene archaeology can plunge in anywhere.

Acknowledgements

We would like to acknowledge the following people for their generosity in helping with material, offprints, advice, reading sections of text and permission to reproduce illustrations: Grenville Astill for references to the east Brittany project; Geoff Bailey for advice on site territorial analysis and allowing us to use the illustration in Box 14.3; David Bescoby for the use of his Malham GIS study; A.E. Bettis III for literature and for permission to publish the figure and tables in Box 10.2 from *Soils in Archaeology* edited by V.T. Holliday and published by the Smithsonian Institution Press, Washington D.C., copyright © 1992; used by permission of the publisher; John Bintliff for literature on his Boeotia survey and wider work in Greece; Geoff Bowden for constant help with computing practicalities; Caitlin Buck for general help and encouragement with GIS and computing matters; Donald Davidson for literature on his Orcadian work; Robert Demaus of The Demaus Partnership for Fig. 9.5; Charlie French for the photomicrograph and explanatory diagram of Figs 10.3 and 10.4 and for valuable advice on micromorphology; his research student, Helen Lewis, for information on her work on the use of the Donneruplund ard; Richard Hingley for discussions on site re-use; John Morgan for printing some of the photographs; Niall Sharples for the Frontispiece and Figs 12.4 and 12.14, and for ongoing discussions, especially about research strategy; Helen Smith for discussions on her South Uist work (Box 14.1); David Wheatley for advice on GIS and for doing the work and drawings on land diversity (Fig. 13.2); and Alasdair Whittle for advice on the Wiltshire case study in chapter 15 and for ongoing discussions over a long period, especially in integrating a diversity of approaches in research. To all we are extremely grateful. We have referred to many people's work and hope we have not misrepresented them too strongly. As is traditional, we also thank our wives, Vivian Evans and Sonia O'Connor, though in the spirit of this book, because we and they are parameters of each other's niches, they are subsumed within its authorship.

Frontispiece. *Chain-sawing a tree on a chambered cairn. The tree has been able to grow in an otherwise treeless land because of the protection from grazing provided to its seedling by the cairn. These knots of preservation are valuable refugia of ecosystems.*

One

INTRODUCTION

Archaeology is primarily about past people and the diversity of cultural adaptations and changes which different peoples and societies have undergone. The raw material of much of archaeology is the remnants of the material culture of past human populations, their buildings, pottery and other artefacts. But the people who manufactured, used and traded the pots, or who worshipped in the temples, were surrounded by a myriad other things, whether tangible artefacts or intangible socially informed perceptions, which modified their activities and behaviour. Above all, people are a part of ecosystems, and the role of the biophysical environment in offering challenges and opportunities to them is fundamental. Charles Elton (1927) memorably defined ecology as 'the study of animals (and plants) in relation to habit and habitat'. So, too, archaeology studies people in relation to habit and habitat, with some studies, such as ours, focusing on the latter, though without ignoring the former. We seek to investigate past biotic and abiotic environments and, in particular, the interactions between them and the human populations which lived in, modified, and were modified by them (e.g. Fig. 1.1).

Figure 1.1 Catastrophic deposition at Garth's Voe, Shetland. A poorly sorted sand is interpreted as a tsunami deposit, probably associated with a submarine landslide in the north Atlantic around 7000 BP. Elsewhere in Scotland, deposits from this tsunami bury Mesolithic sites, showing people and sudden environmental change in close proximity.

Box 1.1 Chronology of developments in archaeology and environmental studies

This is a selective list to show the sequence and coincidence of major events that led to an understanding of the antiquity of people and of the previous existence of environments different from those of the present day.

1715 – A fossil elephant closely associated with a worked stone tool is reported from the River Thames, London. The significance of this find eludes most people.

1717 – Posthumous publication by Michael Mercati (originally drafted in the 1590s) of stone tools previously described as 'thunderbolts'. Mercati describes them as tools from 'before iron was employed for the follies of war'.

1758 – Antoine-Ives Goguet postulates progression through Stone, Copper, Bronze and Iron Ages, a system later adopted by Christian Jörgensen Thomsen and Jens Worsaae.

1778 – Georges Louis Leclerc de Buffon publishes *Époques de la Nature* and lays the foundations of Quaternary biostratinomy.

1797 – John Frere finds stone tools associated with extinct animals at a depth of 4 m at Hoxne, Suffolk.

1811 – De Witt Clinton applies dendrochronology, apparently for the first time, to estimate the age of earthworks in New York State.

1829 – Paul Tournal's work in Dept. de l'Aude shows the close association of caves, people and extinct animals with cut-marks on their bones.

1833 – Henri Reboul defines 'Quaternary' (first used by Jules Desnoyers in 1829) to encompass strata containing the remains of biota which have living counterparts.

1837 – Louis Aggasiz draws together diverse data to propound a theory of past glaciation linked with climatic change.

1846 – Sir Edward Forbes links the 'Glacial Epoch' with the Pleistocene, so making a connection which persists in modern Quaternary studies.

1855 – Alfred Russell Wallace publishes an essay on the mutability of species and links speciation with biogeography. Further, fuller publication by Charles Darwin in 1859.

1856 – Discovery of the eponymous Neanderthal skeleton, variously thought to be an ancient human or a pathological Cossack.

1863 – Thomas Huxley publishes *Evidence of Man's Place in Nature*. John Evans denounces the Moulin-Quignon mandible as a fake.

1864–5 – Edouard Lartet and Henry Christy discover Upper Palaeolithic rock art in the Vézère valley, thus indisputably associating humans and extinct animals. Oswald Heer describes past environmental changes in Switzerland in the context of past climatic change.

1866 – Ernst Haeckel first coins the term 'ecology'.

1867 – Charles Lyell links glacial stages with orbital eccentricities. This is subsequently refined by Milutin Milankovitch (1924), and goes out of favour with nearly everyone, to be reassessed in late 1970s when ocean core sequences show a similar periodicity to Milankovitch's predictions.

1871 – Charles Darwin publishes *Descent of Man*, and becomes truly controversial.

1909 – Albrecht Penck and Eduard Brückner differentiate different stages of glaciation in the European Alps. Several stages of ice advance already described for North America by this time.

1912 – Publication of Piltdown Man, to be formally denounced as a fake in 1951. Gerard de Geer publishes an estimate of the duration of the Post-Glacial period in the Baltic region based on varves.

INTRODUCTION

Chronology of developments in archaeology and environmental studies

This uses small-scale periodicity in sedimentation to date the geomorphological consequences of long-term fluctuations in climate.

1916 – Lennart von Post publishes the first systematic study of ancient pollen, from a peat sequence in southern Sweden.

1925 – Scopes Trial in Tennessee. Anti-evolutionists win the trial, but the publicity works largely to the benefit of pro-evolutionists outside the southern USA.

1927 – Charles Elton redefines 'ecology' and effectively launches a new branch of biology.

1935 – Arthur Tansley introduces the 'ecosystem' concept.

1942 – Robert Lindeman introduces the energy flow model for biotic communities, which becomes the predominant paradigm for the next half-century.

1955 – Willard F. Libby invents radiocarbon dating, and Cesare Emiliani develops oxygen isotope analysis of deep sea cores as a record of past climate.

1960s – Robert H. MacArthur's work on optimality and evolutionarily stable strategies moves behavioural ecology from empiricism to quantified rigour.

Since the 1960s, both archaeology and ecology have undergone numerous changes in outlook and methods. In archaeology, the large-scale 'salvage' excavations in northern Europe and North America led to sampling for 'environmental' materials often with little coherent research strategy. The trend was towards increased descriptive rigour and the characterisation of processes: 'culture' was seen as adaptive behaviour. One major feature of this period was the search for early agricultural sites in the Near East and Mesoamerica. Subsequently, post-processual theory has swung archaeology towards the subjective and the individual, and away from any notion of material culture being a functional response to the environment.

Meanwhile, ecology has progressed in the complexity of the levels of biotic organisation which it seeks to study. The interconnectedness of global ecosystems and concerns over deforestation and global warming have led to ever more elaborate attempts to model complex natural systems. Some long-held ideas, such as that of biotic succession, have been strongly challenged, whilst others, such as Lindeman's energy flow paradigm, have stood up to renewed investigation.

The history of archaeology shows a developing concern with ecology (Box 1.1). Some of the earliest researches concerned stone tools obviously of human manufacture yet found at great depths – and hence of great age – in river gravels and other drift sediments. In some cases, the juxtaposition of such tools with the bones of animals now extinct or of greatly altered geographical distribution confirmed the antiquity of the tools and led to the first studies in the relationship between humans and their biophysical surrounds. Similarly, the discovery of cave art in France and Spain placed humans in an unfamiliar biome (Boule 1923, 252–60). The evolutionary studies of Wallace and Darwin had gone far to destabilise assumptions about the immutability of species, and, along with their biogeographical investigations, made possible the gradual acceptance that plant and animal communities in the past could have been quite different from those of today; equally, humans may have been witness to, if not actually involved in, the changes that occurred (Box 1.1). As these ideas developed in biology and archaeology, geologists were beginning to develop a sedimentary framework for Pleistocene and Holocene (i.e. Post-Glacial) Europe. The gradual resolution of glacial/interglacial cycles provided evidence of major climatic changes, with inevitable consequences in terms of erosion, sea and lake levels, soil histories, and thus of the resources available to human populations at different times and places (de Geer 1910). As long ago as the 1840s, the pioneer prehistorian Worsaae co-operated with geologists and biologists to gain an understanding of the wider environmental setting of past human occupation, and this has been the tradition in Scandinavian archaeology, through the times of de Geer and von Post to the present day. As archaeology grew out of antiquarianism in the early twentieth century, it already had firmly in place an interest in past environmental change on a long-term scale. This growing-together of archaeology and the natural sciences is nicely charted by Trigger (1989, 247–50), who sees in it one of archaeology's numerous 'isms': environmental functionalism.

On a different scale, the remains of plants and animals found on excavated sites of human settlement came to be seen in terms of resource exploitation. The discovery of lake villages in the Alpine hinterland both contributed to the wider consideration of changing lake levels in the Holocene, and provided an opportunity to examine plant macrofossils in particular in a remarkable state of preservation (Keller 1878). Elsewhere, the investigation of crannogs in Ireland led to some of the first systematic accounts of bones from archaeological deposits, written, oddly enough, by Oscar Wilde's father (Wilde 1862).

At an early stage of archaeology's development, then, an interest in the environmental setting of human settlement gradually developed. Initially, large-scale environmental change was used as a tool in establishing relative chronology by interpreting prehistoric deposits against an empirical, biostratigraphic framework. Sequences of macroscopic plant remains – tree stumps, twigs, leaves and seeds – from bogs in southern Scandinavia allowed Blytt and Sernander (Sernander 1908) to establish a sequence of stages for the last 12,000 years which is still used to some extent as a background against which to set more detailed local studies. An important topic in this context was the study of pollen from slowly accumulated peats and lake sediments (von Post 1916). Two key figures in the development of pollen analysis were Sir Harry Godwin (1956), who was principally concerned with long-term biostratigraphy, and the Danish palynologist Johs Iversen (1941) who pioneered the use of pollen as a means of recognising human activity. Since the 1940s, pollen analysis, especially of Holocene deposits, has focused on the interpretation of perturbations in past vegetation cover, sometimes quite minor, in terms of human activity (Simmons and Cundill 1974).

In her preface to a survey of case studies in American archaeology, Patty-Jo Watson argues that the study of past environments has only developed in the Americas since the Second World War (in Reitz *et al.* 1996). Certainly, an overview of the American literature would give this impression, but there are important early exceptions. As early as the 1860s, naturalists exploring the landscapes of North America were beginning to draw attention to the shell-heaps which adorn some parts of the continent, seeing in them both evidence of past Native American activities and information about past biotas (Wyman 1868; Walker 1880; 1885). Archaeology itself had an erratic start in North America. As Fagan (1991, 29–33) points out, it was only in the 1880s and 90s that it came to be generally accepted that the Native Americans were of considerable antiquity in the lands from which Europeans were rapidly ousting them, and that the undeniable sites and monuments which were frequently reported were of Native American construction and not the relics of some other race, unspecified but presumably white-skinned. Trigger (1989, 418) comments on the 'slowness with which archaeology developed in the New World'. This development certainly differed from developments in Europe, not least in an emphasis on the function and technology of artefacts, and hence the frequent use of ethnographic analogy (ibid., 270–5).

As archaeology developed and diversified its approaches to the evidence of the past, the differing influences of disciplines outside archaeology came to bear, particularly Quaternary geology and palaeontology. An influential figure in Europe was Frederick Zeuner, much of whose earlier research was concerned with the stratigraphy of the last million years and with understanding Quaternary deposits on a global scale (Zeuner 1952). Others were influential in bringing this approach into archaeology. During the decades between the First and Second World Wars, for example, Alfred Kennard worked on land and freshwater molluscs in geologically recent deposits as a means of studying environmental change (Kennard 1943). Kennard's often insightful work was influential in the subsequent development of this line of research, but Zeuner's global approach to environment change and geochronology had wider relevance.

In contrast, scholars at Cambridge, especially Grahame Clark and Eric Higgs, were more concerned with the site-by-site interplay between settlement and the availability of resources and, under their tutelage, a generation of Cambridge students developed the field of economic prehistory (Higgs 1972; Clark 1989). Just as Zeuner's approach was broadly biogeographical, so Higgs adapted ideas from settlement geography, such as Site Exploitation Territorial Analysis, a procedure for analysing the resources likely to have been available to a site (pp. 189–91; Box 14.3). Grahame Clark's studies of early European prehistory and its ecological setting (e.g. Clark 1952) were less obviously responsive to contemporary concerns. The great strength of Clark's work lies in its breadth, and in his willingness to tackle big questions of human social and cultural development from an ecological perspective. Clark (1989, 539–59) gives a good account of his intellectual journey, including the early influence of Gradmann and of Harry Godwin, with whom Clark planned some of the first British fieldwork that looked explicitly at the relationships between human populations and the biophysical environment. An important figure in the study of vertebrate resources was one of Zeuner's colleagues, Ian Cornwall, who was instrumental in developing the study of animal bones from settlement sites (Cornwall 1964). Zeuner was succeeded at the University of London by Geoffrey Dimbleby, who wedded an interest in environmental change with contemporary concerns about the environment and the role of people in destabilising ecosystems (Dimbleby 1976). In Europe by the early 1970s, then, this field of research had gone through the stages of borrowing from, and working alongside, other disciplines, and was established within archaeology, being taught as a part of undergraduate courses and routinely involved in excavations and other field investigations.

Developments in North America had taken a rather different turn, in particular through the application of the ideas of Julian Steward. Essentially an anthropologist, Steward was preoccupied with the relationship between culture and environment, calling his field of research 'cultural ecology' (Steward 1955), which postulated that human populations living in similar environments will make similar adaptations to that environment, and that such adaptations will be unstable and responsive to environmental change. Steward's ideas were strongly deterministic, and often deeply flawed, though his approach at least drew American archaeology beyond ceramic and lithic typologies. Not all North American work of the 1950s showed Steward's influence: Waldo Wedel, for example, studied vegetational change on the Plains, to relate subsistence patterns to environmental change, without overt determinism (Wedel 1953).

A key North American figure is Karl Butzer. Although a geographer specialising in sedimentary processes, Butzer developed a holistic, ecological approach to prehistory which managed to accommodate culture change and environmental processes (Butzer 1971; 1982). His rejection of a simple population-pressure model for the development of 'civilisation' in Egypt was based in part on a large-scale analysis of settlement pattern and its relationship with topography and other geographical factors (Butzer 1976), thus linking the environment with social change and the development of a complex society. Butzer's influence has been considerable, not least in his preparedness to tackle the complexity of natural systems in detail, and in his ability to point out the implications which sedimentological data, for example, might have in terms of settlement mobility or resource availability.

From about 1960 onwards, a number of changes were in train within archaeology, altering the ways in which the evidence of the past was studied and regarded. The traditional approach saw the development of human culture as a progression of prehistory to history, with culture succeeding culture, rather as if history progressed towards some predetermined goal, and cultural change was directional. This approach was an almost inevitable product of the Three Age system, which saw prehistory as stages through the use of stone, then bronze, and finally iron, and with urbanisation as some sort of higher development towards which cultural progress aspired. The culture-historical approach to archaeology appealed, ironically enough, both to those who saw human history in terms of striving to realise some divinely inspired destiny, and to those Marxists who saw in the past a rather different, though equally inevitable, progression from barbarism to the eventual triumph of the proletariat (Childe 1936).

It took a surprisingly long time for the culture-historical approach to be questioned, and a remarkably short time for it to fall apart as archaeological theory moved on. During the late 1960s, archaeology moved away from seeing material culture as something shared and passive. The functionalist and processual approaches saw culture more as an adaptive strategy, a means by which people adjust to, and utilise, their environment, and by which they express and modulate internal social pressures and contradictions. Archaeology responded to this change of attitude by developing the study of economic systems and the processes of resource acquisition and utilisation. Particular topics came to the fore, notably the study of early domestication of crops and livestock (Higgs and the Cambridge school again), and the whole question of agriculture as an adaptive strategy (Braidwood 1960; Harris 1989). In other ways, however, the subject remained wedded to an empirical methodology. Excavations were seen as an opportunity to obtain data and to see what those data indicated. If a particular site resulted in the recovery of a lot of charred seeds, then those seeds were deserving of study, regardless of whether the results would be of value to any previously stated research aim. An attitude of 'let's see what we can find . . .' persisted and persists.

On the positive side, an increasing scientific rigour came into the subject, with a much greater concern to ensure that sampling and recovery techniques gave samples of material with known and controlled biases (Clason and Prummel 1977). Much more thought was given to the routes by which the remains of plants and animals came to be buried in archaeological deposits (i.e. the study of their taphonomy), and indeed how those deposits came to be formed in the first place (e.g. Stahl 1996). These developments were made necessary by the advent of large rescue excavations in Western Europe, and of 'salvage archaeology' in North America. Huge sites yielded far more data than could be practically studied, and thought had to be given to predicting which deposits would yield information of high integrity or of particular relevance to specific questions (Gamble 1979; Toll 1988). From the other end of the human time-scale, the investigation of early hominid sites in East Africa also contributed to taphonomic research, in this case in order to establish criteria by which bones accumulated by hominids could be distinguished from those accumulated by other predators or scavengers (Shipman 1981).

During a period of philosophical change in archaeology as a whole, then, there developed a concern with methodology and refining of investigative techniques. One consequence of this has been intensive criticism of the discipline from some archaeological theorists (e.g. J. Thomas 1990). Just as culture-history yielded to functional processualism, so the latter paradigm has been assailed by a post-processualist school who reject the implied determinism of viewing culture as an adaptive phenomenon. Environmental functionalism is deeply unpopular with some post-processual theorists, because it implies that cultural development is constrained by, or at least related to, extrinsic environmental factors, and is not wholly determined by intrinsic social ones.

Studies of past ecosystems have played a key role in giving settlement archaeology a regional perspective. Both in Europe and North America, field archaeology has moved towards investigating the process of settlement and the development of a settled and utilised land (Rossignol and Wandsnider 1992). Excavation opportunities tend to arise site-by-site, particularly in the context of developer-funded resource management on the American model. This tends to focus archaeology on individual sites, rather than on the interrelationship of settlement sites and other forms of land-use across larger areas. Understanding the nature of the areas between the sites may be essential in order to understand the sites themselves, and may also reflect something of the perception which the occupants of those sites had of the land around them. Studies of past environments have thus been central to some of the big landscape archaeology projects of recent years, with the investigation of large-scale environments integrated with that of individual sites.

Archaeology thus studies the relationship between culture, subsistence and environment, but this is no more helpful than saying that geographers study the world. Perhaps geography provides a good analogy, for much environmental research is essentially geographical (Boyd 1990). However, notwithstanding the contribution which Butzer and others have made, the trouble with taking too explicitly geographical an approach is that it plays down questions of cause and motivation. If we are concerned with long-term adaptive strategies for survival and their regional variations and responses to environmental change, then we have to explore the motivation behind the adoption of one strategy rather than another, motivation which may be socially or ecologically informed; and we have to investigate questions of why and how ecosystems change. All of this argues for an ecological approach, concentrating on questions of interaction, whether as the population ecology of people or of a specific crop, or the community ecology of the trophic system of which people are an active part (K.D. Thomas 1989). Reitz *et al.* (1996, 3) have offered the view that our subject 'is directed toward understanding the dynamic relationship between humans and the ecological systems in which they live', which thus focuses on human palaeoecology rather than past human geography. Whether these two terms will eventually be shown to be one and the same and whether either is, in the end, a useful definition for our research are questions to which we shall return at the end of this book. For the moment suffice it to say this book is about archaeology and ecosystems.

Part One

LIFE

'Life' refers to processes which go on in environments and ecosystems, not just life in the biological sense of the word but the flow of nutrients, energy, water and matter, processes which maintain and drive ecosystems and ecosystem change. In chapter 2 we look at domains of the human environment and the interactions of humans at different scales. Chapter 3 reviews biology, soils and sedimentation, and chapter 4 the dynamics of space and time. The base-line for all this is the present day, most of the information being acquired from observations of living or active systems. It will be quickly apparent, however, that it is difficult to keep the perspective of time out of the presentation, and that at least some of the properties of ecosystems are founded in the past.

Two

HUMAN ENVIRONMENTS

This chapter takes an overview of the major external physical and biological factors which define the setting for human activity, providing the challenges and opportunities which may predicate particular cultural responses. This is a brief look at the structure of environments rather than the detail of their constituents. Many geographical texts offer a more detailed review, and Simmons (1997) gives a particularly pleasing synthesis of the interrelationship between culture and environment.

The environment in which we live can usefully be considered as a series of 'spheres'. This subdivision is a matter of convenience rather than a reflection of the workings of the environment, and it allows separate discussion of the atmosphere, hydrosphere, lithosphere and biosphere. In later chapters, we take a more holistic view.

The *atmosphere* is the gaseous envelope which surrounds the Earth in a tenuous, thin layer, and on which living organisms depend. The great majority of it is nitrogen; the largest minority is oxygen. This is an odd composition for a planet in our solar system, and is probably to be explained by the parallel evolution of the atmosphere and the biosphere. Latterly, our own species has had a disproportionate effect on the atmosphere by increasing the partial pressure of carbon dioxide, with consequences which are still a matter of debate. The *hydrosphere* is the Earth's complement of water, whether as oceans, ice-caps, rivers or clouds. The oceans account for about 97.5 per cent by volume of the hydrosphere, the polar ice-caps and montane glacier systems for another 2 per cent, leaving just 0.001 per cent as water vapour within the atmosphere, and less than 0.5 per cent as rivers, lakes and ground water (Mielke 1989, 149). An important function of the hydrosphere is as a 'sink' of carbon in exchange with the lithosphere and atmosphere. The oceans hold around 36,000 billion tonnes of carbon; nearly five times the amount held in the atmosphere, and about seven times the amount represented by the world's terrestrial biota. The *lithosphere* is the Earth's crust, the outer 5–50 km of the mass of rocks and sediments of which the planet is composed, a somewhat distorted sphere of 7×10^{21} tonnes, some 3,964 km in Equatorial radius (Foster 1985). Despite the complexity of crustal geology, only eight elements each constitute more than 1 per cent of the lithosphere by weight, and much the most abundant of these, at 46.6 per cent, is oxygen (Hamblin 1992, 72). Oxygen is thus more than twice as abundant in the lithosphere as it is in the atmosphere. Lastly, there is the *biosphere*, the total mass of all living organisms on the planet, perhaps 20–30 million species including the smart primate with which archaeology is primarily concerned. From a global perspective, archaeology has a rather odd focus. A living biota is essential to the equilibrium of global systems, but humans are not. Indeed, as the biologist E.O. Wilson is fond of pointing out, the planet would carry on regardless if all vertebrate species were to become extinct. Invertebrates and plants are much more important.

LITHOSPHERE

The nature of the rocks and sediments which outcrop at the Earth's surface strongly influence the type of soil which develops, and thus may constrain the development of natural vegetation communities or human crops. *Solid geology* is generally the most unchanging factor in the environment of a particular area. On a human scale, even in terms of the duration to date of our species, the rocks are eternal. In most parts of the world, the solid geology is dated in terms of tens of millions of years at least, several orders of magnitude greater than the human career. Rock outcrops and other visible features of the lithosphere are commonly incorporated into myth and legend, like Ayers Rock, Mount Rushmore and the white cliffs of Dover.

The plates which comprise the lithosphere move around the Earth's surface. Throughout geological time, continents have formed, merged, and split apart, moving and altering their constituent rocks, and altering the distribution of life on the planet (p. 47). The continents are composed of igneous rocks, directly derived from the cooling of magma from relatively deep below the Earth's surface; sedimentary rocks formed by the redeposition of material derived from the weathering of other rocks; and metamorphic rocks, formed by the modification of other rock types through heat and pressure within the crust (Hamblin 1992). The processes of crustal plate movement and vulcanism set up colossal stresses which can lead to the formation of new rocks and minerals, in the latter case particularly when magma is injected into fissured crustal material. As a result, the Earth has a continental solid geology which ranges from large areas of ancient crystalline rocks, such as northern North America and Scandinavia, through thickly layered sediments in coal- and oil-field areas, to highly diverse regions of intense folding, as seen in the European Alps and parts of South-East Asia. The diversity of rock types is patterned spatially, especially at the continental margins, with a decreasing coarseness as one moves from continental interiors to deep oceans (Fig. 2.1). This is an important feature of sediments generally, to which we return (cf. Fig. 4.2).

Overlying the solid geology in many parts of the world is the *drift geology*, a veneer of recent, mostly unconsolidated, sediments (Fig. 9.2). It encompasses alluvium laid down on river floodplains, the clays and sands left behind by wasting ice-sheets, peats, sand dunes, marls and muds (Lowe and Walker 1997, 85–154). Over those parts of Eurasia and North America which were glaciated during the Pleistocene, drift deposits are mostly a few tens of thousands of years old at most. This puts them into the same time frame as human settlement; a river can deposit a significant thickness of alluvium during a human lifetime. Drift deposits such as floodplain alluvium may provide areas of well-watered, fertile soils, thus encouraging dense settlement by agricultural communities. In the past, floodplain agriculture in the Nile valley has largely depended upon management of the deposition of alluvium, making humans an active part of a geological process. Extensive areas of drift, such as the peat of the Boreal tundra or the wind-blown sand of the north-western

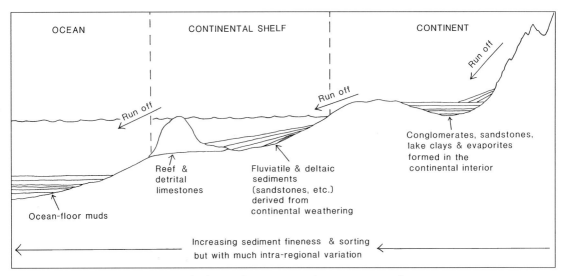

Figure 2.1 Generalised diagram of sediments from continental interior to ocean floor.

European geest, may also obscure considerable variation in the underlying solid geology, thus being important in soil formation.

ATMOSPHERE

Turning to the atmosphere, its most important manifestation for archaeology is climate, which must be distinguished from weather. *Climate* is the consequence of long-term patterns of atmospheric circulation, the movements of warm and cool air masses, variations in pressure and humidity, and surface interactions with the hydrosphere, whether in the form of oceanic water or polar ice (Budyko 1982). The climate of a particular area can usually be broadly predicted by reference to latitude and the proximity to a major ocean, with minor modification as a result of local topography. Patterns of climate around the globe are relatively stable over the long term, although temperate latitudes, in particular, are more susceptible to abrupt changes of climate if parameters of long-term mean temperature and insolation move beyond quite narrow limits. We now know, for example, that northern Europe underwent startlingly rapid changes of climate during the amelioration of the last glaciation, between 15,000 and 10,000 years ago. *Weather*, by contrast, is the day-to-day manifestation of climate, more variable temporally and spatially, and generally of less long-term significance. It is, however, more directly experienced and reacted to by people than is climate. Thus in temperate regions, a temperature variation of 3°C from one day to the next would invite little comment, whereas a continent-wide fall of 3°C in decadal averages would bring catastrophic results in terms of plant growth, snow accumulation and atmospheric circulation. In the long term, it is climate that is of the greater significance in terms of erosion, soil development and biotic productivity, but spells of extreme weather within a period of stable climate may also leave their mark. The denudation of some montane areas in northern Britain has been attributed to the medieval 'Little Ice Age', itself a phase of extreme excursions from the long-term weather averages (Ballantyne 1991), and single storms are increasingly seen as important in soil erosion and increasingly being identified in the archaeological record (Bell and Boardman 1992).

Topography

Climate, weather and geology interact to produce the topography. Erosion by wind, ice and water reduce relief and move material from areas of higher relief to areas of deposition. Some topographical features reflect geological structure, such as the massive mountain chains produced where continental crustal plates collide, though even these may be greatly reduced with time, as with the Appalachians. The relief of the Himalaya-Karakorum-Pamir region results from the more recent collision of the Indian and Asian plates, but even here, it is relatively recent erosion by rivers and glaciers which has produced the detail. The deep gorges which result lend themselves to cultivation by terraced fields, hence the characteristic patterns of settlement. Over much of the planet, most topography can be dated to the last million years, and in some areas to the last few tens of millennia. In any region, the topography is a palimpsest of different periods of crustal movement and erosion, with the most recent effects often being the most evident. Latterly, humans have become significant geomorphological agents, and there are few landscapes on Earth today which do not show some human influence.

Biosphere

The biosphere of which we are a part can affect topography, and has significant interactions with the atmosphere, hydrosphere and lithosphere. Green plants are important in maintaining the chemistry of the atmosphere, and significant in terms of the weathering of rocks, for example, by oxidation (Sparks 1986, 22–48). Terrestrial plants have a physical and chemical role in topographical development, whether by causing weathering and erosion on rock outcrops by root action and chemical secretion, or by inhibiting the movement of weathered material by holding it within a root system. Animals are also intimately involved with the lithosphere, creating it with their carbonate secretions as with coral reefs, influencing groundwater drainage through burrows, and cycling carbon and nitrogen between the biosphere, atmosphere and lithosphere (Mielke 1989, 41–57).

The medium in which biosphere, lithosphere, hydrosphere and atmosphere come into their closest and perhaps most complex interaction is in soil. This is a living medium, the product of the influence of air and water on the geology, and containing and sustaining a biotic community which in turn modifies its physical and chemical properties. There can be a fine distinction between a soil and a sediment (Box 10.1), but perhaps the clearest division is that a sediment lacks the stability of a soil and the minero-organic compounds that are a consequence of this stability. We discuss soils and sediments at greater length in chapter 3.

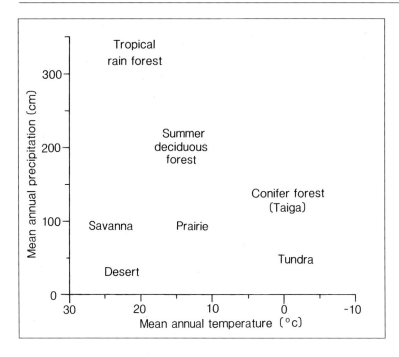

Figure 2.2 Relationship of some major plant groupings to climate at the present day. For archaeology, this is an unsatisfactory classification because groupings of plants will have been different in the past.

Species are not evenly distributed across the surface of the planet, and the study of their distribution is termed *biogeography*. In any one continental land mass, major distributions can be shown to relate to climate (Fig. 2.2), but the situation with regard to particular species is often more complex. The freshwater biologist Macan provided a useful basis for considering distributions by suggesting that the starting question should be 'Why doesn't x live here?' (Macan 1963). Possible answers would have to include lack of adaptation to particular conditions, exclusion because of competition with a better-adapted taxon, or the presence of a geographical barrier. Mielke (1989, 96) draws attention to the greater north–south variation in biota in Eurasia, where the major mountain chains are mostly aligned across north–south migration routes, than in North America, where the Rockies and Appalachians form more of a barrier to east–west movement. Students in both biogeography and archaeology have been slow to appreciate the links between the two fields, especially in terms of time and history. Biogeography allows an understanding of past distributions of species, many of them exploited by or competing with humans, while archaeology can provide the dimension of time for modern distributions (Deevey 1949; Colinvaux 1993, 338–63).

Many plants and animals modify their immediate environment, creating conditions favourable to their life, reproduction and dispersal. Termites build mounds, beavers dam rivers, and deciduous trees maintain humus-rich soils by their annual leaf-fall. Humans are unusual in their ability to modify both their environment and their behaviour. The beaver has been highly successful in colonising northern temperate latitudes, but its dam- and lodge-building activities are not a successful adaptation to high polar latitudes or the tropics, and beavers have not developed alternative ways of ensuring population continuity in these different environments. They are thus restricted in their global distribution. Other animals, such as foxes and rats, have shown an ingenious ability to adapt to new opportunities and challenges, yet they still lack the capacity to make innovative modifications to their environment which would allow them to spread still further. Rats spread by hitching a ride with humans, but none the less rely on being transported to a suitable environment in which to do essentially what rats do elsewhere in the world. Humans have succeeded in colonising the planet precisely because of their capacity to modify their behaviour to adapt to the challenges and opportunities presented by a new environment, together with the ability to bring about local environmental change to facilitate a particular cultural adaptation.

People

The distribution of humans is thus in part related to climate and physical geography and in part to the adaptations of social and economic behaviour and material culture. Archaeology seeks to examine this interaction as the key to the cultural diversity of humans, and to the nature of much of the terrestrial environment as we experience it today. This last phrase raises an important point. Human populations respond to the environment as they perceive it (Kirk 1989) and this can vary with the individual, the community or the culture according to, for example, tradition, recent events, technology and social aspirations. The perceived environment will be a subset of the operational environment within which that particular population exists, although it can also extend beyond it. The operational environment will in turn be a subset of the geographical (or objective) environment as a detached observer might describe it (Butzer 1982, 252) (Fig. 2.3).

It is a moot point whether we can describe the geographical environment even for the modern world, as our description will be coloured by our own perceptions and prejudices. The geographical environment can remain the same while people's attitudes to it make differences, as in different attitudes to marginal land (p. 56). An account of whale populations in the Antarctic today would probably be written in terms of our obligation to conserve these gentle, threatened creatures, which cling to existence in a last refuge from people. An earlier generation, on the other hand, would have discussed whales as a valuable resource and regarded the inaccessibility of the Antarctic not as a refuge for the whales but as a threat to whalers (e.g. Ommanney 1938).

Cultural relativism is important. In archaeology, we depend upon those attributes of past environments which can be discerned from the evidence. If we work principally on material from sites of human occupation or other activity, then we are investigating the past operational environment and there will be aspects of the contemporaneous geographical environment which will not be apparent. Ultimately, though, if we wish to understand the interaction of people and environment, it is the past perceived environment which is important and there would seem to be no way in which that can be examined. However, such human intervention in the past environment as is seen in the archaeological record is itself a reflection of past perceptions and, as Trigger (1989) points out, if perceived environment and operational environment were too different, the human

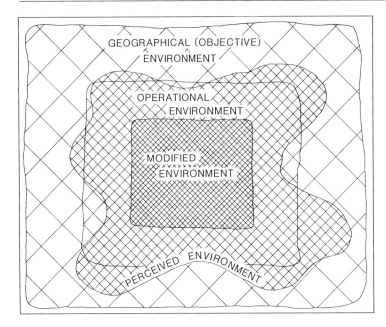

Figure 2.3 Categories of environment in relation to people. Archaeological sites tend to be in the modified environment, although offsite studies also sample the operational environment.

population would fail. We can proceed with some confidence, then, but must be aware too of the emplacement of our own perceptions onto the data. For example, during the later Bronze Age in parts of northern Europe there is some evidence of a fall in mean temperatures and an increase in precipitation. To modern eyes, from the perception of an industrialised people dependent upon intensive cereal agriculture, this is seen as a climatic 'deterioration'. However, whether contemporary peoples saw it as such, or noticed it at all, is uncertain. When discussing past climates and environments, therefore, we need to be specific about parameters such as length of growing season and altitudinal limits on plants, and to avoid subjective terms such as 'deterioration' unless we can be sure that this is how such change was perceived.

The human environment is complex, difficult to subdivide and defined in terms of contemporary perceptions, an issue to which we return in chapter 16. This book takes an overtly ecological view of past human populations, and the next chapter considers in more detail some of the concepts from modern ecology and biogeography which are useful when examining the environment of past human populations.

Three

Ecology and Environment

This chapter is about components of environments preserved in the archaeological record as material remains, such as plants, animals, soils and sediments. Only if we can understand the processes responsible for present-day systems can we interpret ancient remains.

The Biosphere

Species

An important concept in biology is the species. Members of a species breed with each other, but not with members of other species, and generally interact far more frequently with other individuals of their own species. Globally, a species usually exists as a number of related populations, often geographically separated, within each of which not every individual breeds. Some are too young or too old and some, like worker bees, have other functions. With the exception of those formed by the splitting of a fertilised egg or by asexual reproduction, no two individuals are genetically identical, nor are any two individuals identical in morphology or behaviour. As a result, some individuals will be more successful in obtaining food, shelter and mates, and so more likely to contribute young to the next generation. These more successful individuals are better suited, or better fitted, to the environment in which their population lives. This is natural selection as described by Darwin and it maintains life in the groups of interbreeding individuals that we call species. Every species ultimately becomes extinct. Species in the geological record therefore have a time depth and are called *chronospecies*. It is impossible to check their breeding viability with present-day species: identical morphology is often taken to indicate that a chronospecies is identical with a modern species. But other characteristics such as calls, chemical emissions, behaviour and physiology, which are not preserved fossil, can separate species which are anatomically close or identical. Increasingly, biochemical and serological assays are used to separate closely related species. Fragments of DNA are retrievable from bones and other biological remains of up to a few millennia old, and can be used, for example, to gain a little information about the genetic characteristics of a mammoth alive at the end of the last Ice Age (Hagelberg *et al.* 1994).

Species' ranges

The range of a species is related to environment, each species requiring particular conditions for reproduction, feeding and general activity. Species of great antiquity may have been widely dispersed by movements of the continental crust, though it is probable that such widely separated populations will generally have diverged to separate species. Otherwise, species are confined to the biogeographical provinces in which they evolved, as defined by major geographical barriers to dispersal. Only hominids, and the animals and plants they have transported with them, have repeatedly overcome geographical barriers. Species do not always occupy the whole of their potential range, however, especially after a major environmental disturbance, when some species may spread slowly into the new environments. Rates and mechanisms of dispersal also play a part. Land snails disperse slowly, so the land-snail fauna of Europe has many small total-species populations, especially in the south. Lake molluscs, in contrast, disperse rapidly because of their contacts with, and transport by, waterfowl, and so most species are widespread across Europe. We may also see relict distributions, where a formerly widespread range has been split up by climate or other environmental change into only a few widely separated localities. Overall, ranges can be cosmopolitan, related to biogeographical provinces, regional or locally endemic, or in disequilibrium where there is a time lag between environmental change and response. This all tells us something about evolutionary origin, environmental change and method and rate of dispersal; ranges and distributions are not only related to the contemporary environment but reflect past conditions and the history of the species.

Within-range variation

Within a species' range there are separate breeding populations, which have evolved in response to different environments and have their own gene pool, behaviour, ecology and, sometimes, morphology. Within some species, these groups may be so different from each other that they are seen as subspecies; at a smaller spatial scale, there are races or strains; locally, the term *deme* or *panmictic unit* is used for populations of individuals which regularly interbreed (Mayr 1970). For example, in the case of cornbuntings (*Miliaria calandra*, a small seed-eating bird) in southern Britain, demes are small – only a few square kilometres – and not related to obvious environmental features. Even so they are separate breeding units, kept apart by subtle differences of song. Some species of land snail have panmictic units only a few metres across. Again there is an embeddedness of time. Categories covering larger areas, such as the subspecies, are older and have less genetic cohesion or gene flow than those covering smaller areas like the deme. This is the basis for using DNA and other biochemical studies as indicators of the timing of past movements and evolution.

Each population occurs in a *habitat*, characterised by particular biological, chemical and physical properties, which varies across its range as related to climatic gradients (Fig. 3.1 and Table 3.1). In mid-range where climate is most suitable, a species is *eurytopic*, i.e. living in a variety of habitats and tolerating hostile conditions such as reduced shelter from predators, scarce food and high parasite infestation. It may also be genetically diverse and, therefore, able to respond flexibly to change. This would be the situation in A3 and B2 in Fig. 3.1. Nearer the edge of its range where climate is limiting, and where, accordingly, more congenial conditions are needed to counterbalance its adverse effects, the same species may be *stenotopic*, i.e. living in a reduced variety of habitats. This would be the situation in A4, B1 and B4 in Fig. 3.1. However, a habitat that is harsh physically may be congenial through the exclusion of competitors. Populations at the edge of their range are often of limited genetic diversity; but they can be remarkably permanent. We see this in pests of human products like grain beetles which exist as isolated populations over their entire range, each often established by very few individuals. This founder effect is important as it results in a population with initially low genetic diversity, and frequencies of genetic traits which may differ markedly from the population from which the founder individuals came. Viewing all habitats together over its range, a species is usually eurytopic, but it is the ecology of a single population in one locality that is crucial in studies of past ecosystems, and here a species may be stenotopic or eurytopic. Inferences about local ecology cannot be made without a knowledge of the regional climate or other general factors like human land-use or associated species (Fig. 3.1). Sometimes

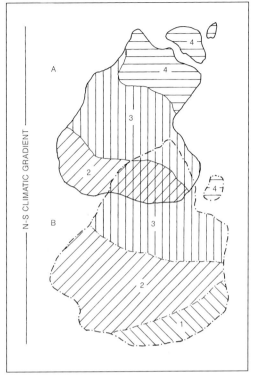

Figure 3.1 Two different total ranges, A and B, of the same species, in two different periods of climatic conditions, showing different ecologies, 1–4, accordingly. Note the disjunct distributions of 4 at the northerly extremes.

there is *polymorphism*, where two or more morphs of a species occur in the same breeding population, and sometimes there is a known relationship between environment and morph abundance, like the camouflage colour and banding of the snail *Cepaea nemoralis*. But this does not occur in every species and even when it does it is not always certain what the relationship is between morph and environment.

Two points of detail emerge. First, demes or other geographical variants are not completely genetically isolated; if they were they would evolve into separate species. Some individuals, often young that have yet to breed, are more mobile than the majority of the population, and spread outside it. Genetic contact takes place, gene flow between demes is maintained, and so is the integrity of the species. Second, breeding populations can be physically mixed but genetically separate. This is the case with domesticated and wild forms of the same species where differences are maintained by human partitioning, or when an invading population replaces a resident one, or during the early stages of succession (pp. 51–2).

In contrast to the above variations which are created largely by natural selection, variation in size, morphology and colour can also occur in response to environment as an individual

Box 3.1 r- and K-strategists

We have used these terms with reference to different reproductive strategies and the wider question of population growth and the idea that this may be regulated by population density. If we model the growth of the population of any species in a closed system, we can show that the rate of increase of population is a function of the starting population size (N), the intrinsic rate of increase of the species concerned (r), and the maximum equilibrium population number which the system will support (K):

$$\frac{dN}{dT} = rN(1 - N/K)$$

This is the 'logistic equation'. Population ecologists accept that it is a gross over-simplification of population increase, but it is a satisfactory model to which many populations can be shown to conform. As N approaches K, then N/K approaches 1, and the right-hand term in the equation tends towards zero. In other words, as the carrying capacity of the system is approached, the rate of population increase declines towards zero. On this model, a plot of the rate of population increase through time is a parabola which first rises, then peaks, then falls.

r-strategists maximise their evolutionary fitness by having the capacity to reproduce rapidly in an uncrowded environment. Typically they show early maturity, have large numbers of small young, are often small in size and tend to have quite short lives, of which a large part is devoted to reproduction. In an environment which is uncrowded, such as a newly created clearing around a human settlement, r-strategists have the opportunity to increase their numbers rapidly, at least until N approaches K, or until some environmental perturbation wipes most of them out.

K-strategists, in contrast, maintain their populations close to the carrying capacity K. Typically, they show late maturity, have few, but large, young, are long-lived and put less of their lifetime physiological effort into reproduction. K-strategists are more successful than r-strategists in relatively constant, stable environments, in which populations have time to increase to values of N close to K, and the advantage lies with those species which can maintain high values of N, rather than with those which can rapidly attain them.

Oceanic islands generally offer relatively stable environments, in which K-strategists such as albatross species, *Diomedea* spp., are successful. However, archipelagoes offer some birds the opportunity to colonise where numbers of other bird species are low (Diamond 1973), and an r-selected opportunist such as the ashy monarch butterfly (*Monarcha cinerascens*) may increase its numbers very rapidly, only to have its population growth curtailed as later, more K-selected, colonisers arrive.

In some cases, one and the same species may be an r- or K-strategist according to the conditions. In these, the ecophenotype may vary accordingly, as in the freshwater snail *Lymnaea peregra* (Calow 1981; Lam and Calow 1988).

Differences between age structure and ecophenotype of adults of Lymnaea peregra *as related to different environments in the same species*

Harsh, exposed environments, e.g. wave-swept lake shores and fast-flowing streams = lotic environments	Mild, sheltered environments e.g. ponds and slow-flowing canals = lentic environments
Growth to breeding rapid	Growth to breeding slow
Many young, few survive	Few young, most survive
Adult ecophenotype, small, with large aperture	Adult ecophenotype, large, with small aperture

ECOLOGY AND ENVIRONMENT

r- and K-strategists

Lentic forms   Lotic forms

Different ecophenotypes of Lymnaea peregra *from lentic and lotic habtats (see Table). Aperture size is related to foot size, individuals living in flowing water needing bigger feet to withstand water current (Lam and Calow 1988).*

develops, the different forms being called *ecophenotypes*. When this is discontinuous, the phenomenon is known as *polyphenism*, to distinguish it from genetically based polymorphism. Limpets (*Patella vulgata*) show this well, with individuals low on the shore where wave action is minimal being less conical than those higher up where it is more severe. Variations in abundance of young and rate of development are also related to environment (Box 3.1 and Table 3.1). Some species are adapted to changing or hostile environments by putting their main reproductive effort into producing large numbers of young with the advantage that at least a few will become adults (opportunist species, or *r-strategists*). Others, living in less challenging or longer-lasting environments, have fewer young and put more effort into developing all to maturity (equilibrium species, or *K-strategists*).

Across-species variation

Size, shape, colour or body texture can be used to group individuals irrespective of species. Cain (1977; 1978) showed two main forms of shell shape among many unrelated taxa of snails, adults being either tall and narrow or short and flat, with the former favouring steep surfaces, the latter gentler gradients or horizontal sites (Fig. 3.2). Globular forms are largely confined to endemic species of tropical rain forest in isolated archipelagos.

Mobility and aggregation

No species is totally sessile. The adult stages of most plants, except for some unicellular algae, are fixed, but mobility is achieved through dispersal of fruits and seeds. Many

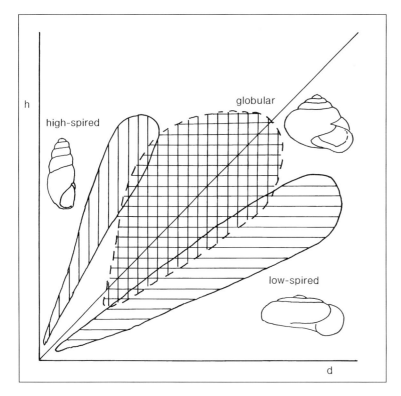

Figure 3.2 Distribution of shell shape in various gastropod (snail) groups; h = shell height, d = shell diameter, oblique line is reference guide to totally globular shells. The separation into high-spired and low-spired forms is typical for most groups, marine and land, fossil and modern; the globular group occurs mainly in oceanic island woodlands. (After Cain 1977, 1978.)

sessile animals, e.g. tapeworms, barnacles and corals, have fixed adult stages but are dispersed through the dissemination of eggs and larvae. Other species are more mobile in their adult stages, but there is much variation. Land snails, and many insects, do not move more than a few tens of metres. Others, like many small mammals and birds, move around more widely, but still within a few hectares. Yet others feed over several tens of square kilometres and, in addition, undertake long-distance annual migrations, e.g. many of the social ungulates like cattle, antelopes and deer.

In terms of food-getting strategies, many species, especially carnivores and woodland species, feed solitarily or in small groups, and partition the environment in terms of day- or night-time feeding. In contrast, open plains animals, especially the herbivores, feed in large aggregations, probably as defence against predators, and are generally active only during the day. But a single species can be broad or narrow in its feeding, varying throughout the year, especially in association with the breeding cycle. Predatory birds and mammals focus on high protein food when they are feeding their young, and are dispersed and territorial, but scavenge carcasses or eat plant foods outside the breeding season and may then be aggregated to allow easier searching. Roe deer (*Capreolus capreolus*) utilise a wide range of plant foods through the year but select particular species seasonally to obtain an energy-efficient diet high in sugar and low in fibre (Tixier *et al.* 1997). Aggregated feeding is also related to food patches. Not all the food is taken, it being more effective to find a new source than to deplete one that is giving increasingly poor returns.

This is also beneficial to the food species, in allowing it to recover and in reducing intraspecific competition.

In temperate regions, seabirds nesting on rocky coastlines are zoned by species (Box 3.2). Movement away from this pattern occurs on a daily cycle (circadially), with areas for feeding (the sea for shearwaters (*Puffinus* spp.), inland rubbish dumps for gulls (Laridae)), preening and washing (freshwater) and display (around the nest area), each at a particular time of day or night. Seasonal change relates to the breeding cycle. Many seabird species congregate in the temperate zone in late winter to early summer for breeding and during the rest of the year disperse out to sea or, in the case of shearwaters and the Arctic tern (*Sterna paradisea*), migrate to the southern hemisphere. Waterfowl such as ducks and geese (Anatidae) show an opposite pattern, aggregating in winter on temperate latitude mudflats and estuaries, but dispersing during the summer in the higher latitudes where they breed.

Niches

There is never complete habitat overlap of the populations of two or more species. The species of insectivorous birds in a woodland partition food by feeding in different areas – from leaves, from bark, in mid-air as with flycatchers (Muscicapidae), and at different heights. All use the same general category of food, but different strategies allow many species to be accommodated in the same area of woodland. The habitat can thus be described in functional as well as physical terms, and the term niche (Box 3.3) is applied to this broader view – how a species gets its food and the range it feeds on, where and when it breeds, and its relationships to other species and physical factors.

Abundance

Why do species occur in different abundances, and why do species' abundances not increase indefinitely? The general level of abundance of a species relates to its position in

Table 3.1 *Differences in one species between the edge and middle of its range*

Edge	**Middle**
Low abundance	High abundance
? Large breeding units, loosely packed	? Small breeding units, densely packed
Limited genetic diversity	High genetic diversity
Potential for inbreeding	Less potential for inbreeding
Limited response to environmental change	Flexible response to environmental change
Low variety of habitats and narrow niches; stenotopes	High variety of habitats and broad niches; eurytopes
r-strategists	K-strategists
Density-independent and physical factors important	Density-dependent and biotic factors important
Susceptible to local extinction	Less susceptible to local extinction

ENVIRONMENTAL ARCHAEOLOGY

> **Box 3.2 Temporo-spatial variations in seabirds and waterfowl**
>
> The table shows the daily to annual changes in seabird and waterfowl populations at different spatial scales. Annually, there are migrations from lowland grassland during autumn and winter to upland or high latitude regions in spring and early summer. 'Zonation' refers to the situation where there are strong aggregations of separate species.
>
> *Circadially*
> Specific areas for feeding, washing/preening, resting/ nesting, and display/courtship
>
> | | *Annually* Late winter/early summer | Late summer/late winter |
> |---|---|---|
> | Seabirds | Aggregated, breeding on cliffs and rocky islands | Dispersed out to sea |
> | | Strong zonation | Weak zonation |
> | Waterfowl | Dispersed, breeding in tundra areas | Aggregated, in estuaries and mudflats |
> | | Weak zonation | Strong zonation |
> | Manx shearwaters and Arctic terns | Northern hemisphere | Southern hemisphere |
>
> *Inter-generational*
> Population fluctuations

the food/energy pyramid (Fig. 3.3), the width of its niche and its life-cycle strategy (Box 3.1). Any one of these can change, either during the year or on a longer term, so even at this basic level there can be variation within a single species. Beyond this there are three kinds of abundance variation. First it can be circadial, as part of behaviour and feeding, and annual (sometimes longer) as a part of the life-cycle (Box 3.2). Then there is variation over a few years, often as a response to predator–prey interactions. Third there are variations over much longer periods of time – decades to millennia – in relation to external change.

The second category is of most interest to neoecologists. Abundances can relate to extrinsic factors which can be biotic, like parasitism, predation and disease, or physical like weather and flooding, or to intrinsic factors such as competition for food and space. In simple ecosystems, relationships are indicated by cyclical changes of predator and prey oscillating together but with a slight time-lag (Elton 1927). In some cases these factors are density-dependent: the effects become proportionally greater as the population increases. Nicholson (1933; 1954) suggested that numbers are regulated in this way, especially where competition for space reduces breeding success. Lack (1954) suggested that there were maximum densities in some bird populations, always rapidly achieved after catastrophic population drop, but never exceeded. Importantly, density dependency can be self-

ECOLOGY AND ENVIRONMENT

Temporo-spatial variations in seabirds and waterfowl

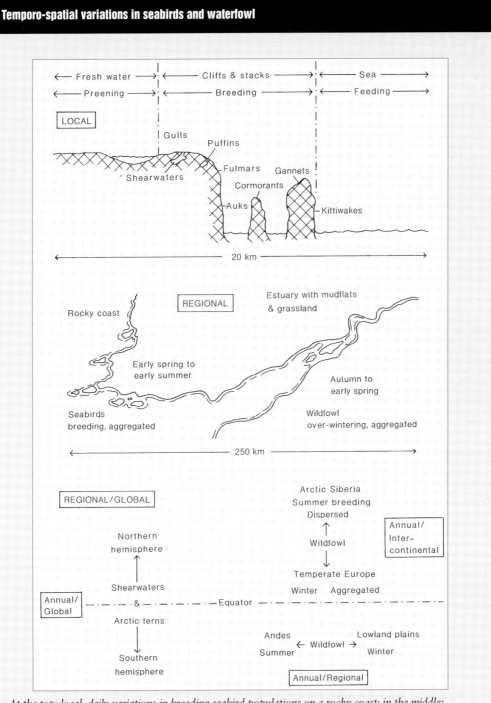

At the top: local, daily variations in breeding seabird populations on a rocky coast; in the middle: regional, annual variations in seabird and wildfowl populations in an area of cliffs, rocky offshore islands and extensive estuarine mudflats and saltmarsh; at the bottom: intra-continental and global annual changes in seabird and wildfowl populations.

regulatory, involving nothing more than the activities of the organisms themselves and the area (or volume) available to them. Examples are numerous: the production of toxins in microbial populations (a problem in germ and biochemical warfare, and brewing technology); and the senescence of vegetations, especially grasslands, which experience cycles of decline and recovery without the external influence of overgrazing or climate change. An alternative view, based on insect populations, is that there is no limitation through density dependence, populations increasing until the carrying capacity of the environment is exceeded or an extrinsic event like climate change intervenes (Andrewartha and Birch 1954). Importantly, a single species may be controlled by either mechanism according to location in the species' range, with biotic and density-dependent factors operating in the middle, physical and density-independent factors towards the edge (Krebs 1985) (Table 3.1).

Multiple-species communities

Relationships between species may take the form of competition for food, breeding areas, or shelter, or may be more direct, as between herbivores and their food plants, predators and their prey, or parasites and their hosts. Species can be categorised in *trophic levels* defined by food sources and the ways in which it is converted to energy and new biological matter. At the lowest level are the primary producers, the green plants; then come the primary consumers, the herbivores; followed by a succession of carnivore levels to the final consumers or top carnivores. Because the amount of energy which each trophic level can crop from the one below is only a small percentage of the energy cropped by that lower level from the one lower still, there is a sharp reduction in energy transfer and in animal and plant biomass. This is a *biomass* or *energy pyramid* (Fig. 3.3), and is one of the reasons why plants at the bottom are common while large fierce animals (usually carnivores) at the top are rare (Colinvaux 1980). Decomposers and scavengers cut across the pyramid. They include organisms as diverse as hyenas, vultures, mackerel, dung beetles, soil fauna, fungi and bacteria, all of which dispose of dead organic matter at all

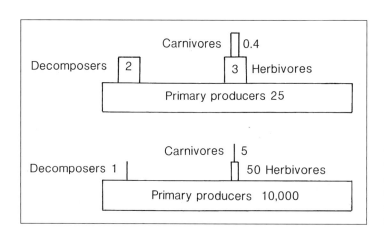

Figure 3.3 Pyramids of energy in Kcal/m^2/day (top) and biomass in g/m^2 (bottom) of plants and animals in immature summer deciduous woodland.

Box 3.3 What do we mean by 'niche'?

The term 'niche' is widely used, not least in this book. It is important to understand that the term is used in at least three subtly different ways by ecologists. These three definitions are closely linked, but the differences between them are important from our point of view.

Eltonian niche

Sometimes described as a Class 1 niche, this is the definition of the term as originally proposed by Charles Elton (1927). The niche of a species is defined by its function within a community. Elton used to explain this definition by saying that when he saw a badger during his morning stroll through the woods near his home, he would think 'There goes a badger' in much the same way as one might think 'There goes a vicar', with all the implications of the role which that individual has in the general community of that place and time. The Eltonian niche is thus an animal's place in the biosphere. Comparable niches in the ecosystems of different places – the top carnivore niche, for example – may be filled by ecologically equivalent species, in this case tigers and polar bears.

Niche as part of the species definition

This definition pre-dates Elton, going back to the work of Grinnel (1904), though most clearly defined by Colinvaux (1982). In this Class 2 niche, the niche defines the function of a population, but so narrowly that only individuals of one species can occupy that niche. The niche is effectively defined by a specific set of capabilities characteristic of one species. Although of some value in neoecology, this definition is rather too narrow for work in palaeoecology, where we may be able to postulate the community function of a given species, but are unlikely to be equally confident about the full range of capabilities of one species population as distinct from another which under a Class 1 definition would be ecologically equivalent. Thus we can discuss the niche of the extinct sabre-tooth cat, *Smilodon*, in Eltonian terms, but not with the detail and precision required by the Class 2 definition.

Niche defined by environment

The Class 3 niche was concisely defined by Hutchinson (1978) as 'a multi-dimensional hypervolume of resource axes', a definition which requires explanation. The niche of a species can be defined in terms of the environmental 'space' in which that niche is carried out. Thus the Class 3 niche of a species is the set of conditions within which the species can live, reproduce and be biotically successful. This is a quite different paradigm from that which underlies Class 1 and 2 definitions. The Class 3 definition describes the coincidence of resources which the species needs, rather than some attribute of the species' physiology and ethology. This is a useful definition for some purposes, not least because the Class 3 niche of an organism can be represented graphically as a volume bounded by limiting values of particular resources. The distinction between the *fundamental* and *realised* niches of a species can then be seen as a reduction of the resource hypervolume by the narrowing of the limiting values of one or more of the resource axes which delimit the niche.

Each of these definitions has its strengths and weaknesses. When discussing past communities, we tend to think in terms of the Eltonian definition. For analytical purposes, though, the Hutchinsonian model is valuable; it allows us to explore the effect of a specific environmental perturbation on the fitness of one or more species, thus allowing the testing of models and hypotheses. The Hutchinsonian model allows us to define 'vacant' niches, i.e. ones which no species currently occupies. Some ecologists take the view that a niche is defined and identifiable only when a species occupies it. We prefer this view, as it defines the niche by what species do, rather than by what they might do.

trophic levels (Fig. 3.3). Omnivores, including humans, do much the same. The success of early hominids may be attributable to their scavenging and/or omnivore niche, allowing them a greater degree of choice than the more narrow niches of associated frugivore or herbivore apes or competing carnivores.

Trophic levels are bonded by the flow of matter and energy, the subject of bioenergetics. This is about efficiency and rates (i.e. there is a time component), useful indicators of which are productivity and biomass. *Productivity* refers to the rate of production of biological material (weight per unit volume per unit time). Primary productivity is the production of plant material; secondary productivity the production of animal material. *Biomass* refers to all living and dead organic matter in a given study area. Productivity and biomass are linked. In a wood, for example, there is the total biomass, which includes all the timber, roots, twigs and leaves, and there is the new biomass, which is the new growth produced each year.

Species occur as *communities*, combinations of particular abundances which repeat themselves in particular habitats. Plant ecologists classify and name communities after the predominant species – the subject of *phytosociology* (p. 178). However, there is a degree of chance in species compositions, related to the different rates at which species become established after environmental change, the environment before establishment, and the fact that species spreading into an area may be able to co-exist where they could not do so under less changing conditions. These are known as *disequilibrium communities* and in humanly modified ecosystems they may be the norm.

Classification of the biosphere

Classification of the biosphere into communities is traditional in neoecology, particularly where these relate to particular environments of, for example, climate, hydrology and vegetation (Fig. 2.2). However, as a point of reference for palaeoecology, this is unsatisfactory since it excludes additional past possibilities, especially the disequilibrium communities described above. One way round this is to group organisms irrespective of present-day species and community ecologies, while not abandoning either the species or community concepts. Properties of biological assemblages (in a given volume) from which community characteristics may be inferred are:

1. Numbers of individuals indicate productivity.

2. Numbers of species, or inventory diversity (Magurran 1988) indicate bioenergetic, competitive and structural complexity, general suitability for living, age and area size. Low numbers imply simple, extreme, young or small communities; high numbers imply complex, equable, old or large ones.

3. *Equitability* or evenness (chapter 13) gives the same information but is more complex in combining species numbers and numbers of individuals of each species. It is expressed as a rank-order curve (Fig. 13.1), simple communities like grassland, saltmarsh and peat bog having L-shaped curves with only one or two main species and a few subordinate ones (low equitability), more complex ones, say of fen wood or rocky sea-shore, having a gentle

curve with a greater number of species and without the predominance of a few (high equitability).

4. Biomass is a better measure of productivity than numbers of individuals and a better measure of inventory diversity and evenness than one based on numbers of individuals and species because it takes into account size differences between species and age groups.

5. Trophic levels, with biomass, indicate food webs and bioenergetic complexity. It is not always necessary to know individual ecologies because particular trophic levels are often a feature of broad taxonomic groups (e.g. hoofed mammals are mostly herbivores).

6. Life-cycle stages, age and sex ratios and ecophenotypes indicate breeding vs. non-breeding communities, seasonality, breeding strategies, and hence community equability or permanence.

7. Intra-specific morphs and ecophenotypes, and cross-specific features such as colour, indicate community properties in relation to their adaptiveness. Extinct species may be assigned to particular ecologies on the basis of their affiliations to higher taxonomic groups. Thus European snails of the family Clausiliidae almost all live in mature woodland, often on tree trunks, or on damp rocks or scree. An extinct species of this family would be unlikely to be a species of dry grassland, at least unless there was strong evidence to that effect.

From these data, we can infer fundamental characteristics of plant and animal communities, all of which can be obtained without knowledge of individual species' ecologies. These are: age since the last major environmental disruption; state of equilibrium; bioenergetic complexity, that is the number of trophic levels and food-web links; productivity; biomass and its spatio-temporal distribution annually; and structural complexity.

SOIL PROCESSES

Soil is the interface between the lithosphere and atmosphere or hydrosphere (chapter 2). It has a mineral component, ranging from gravel to clay (Table 3.3), combined with organic humus to form organo-mineral complexes, which in turn occur as the crumbs which give soil its structure. The mineral and organic compounds influence soil chemistry, whereas the *crumb structure* affects drainage and gaseous diffusion and provides surfaces and spaces for chemical and biological activity. Soil composition, structure and processes are influenced by the wide range of life which soils contain.

Soil formation

Soil forms initially by *weathering* of parent material. Physical weathering includes cycles of heating and cooling, and of wetting and drying, which can bring about the disintegration of rock surfaces; crystallisation, when the formation of crystals of ice or soluble minerals like gypsum expands cracks in the substratum thus exposing fresh

ENVIRONMENTAL ARCHAEOLOGY

> **Box 3.4 Soil horizons and soil types**
>
> ## Soil horizons
>
> The classification is taken from Avery (1973) and Burnham (1980). The horizons are listed in order from the surface, downwards.
>
> *O horizons*: surface organic material, principally leaf litter and other plant remains (O = organic)
>
> Of: mainly fibrous peat
> Om: mainly semi-fibrous peat (m = mor or raw humus)
> Oh: mainly amorphous organic material, uncultivated (h = humus)
> Op: amorphous organic material, mixed by cultivation (p = ploughed)
>
> *A horizons*: the 'topsoil', in which humus and minerals are mixed; usually the most biologically active horizon
>
> Ah: uncultivated, with more than 1% organic matter
> Ap: mixed by cultivation
> Ag: partially gleyed, i.e. intermittently waterlogged, often with rust colours along root channels and cracks
>
> *E horizons*: often pale mineral horizons from which iron oxides, humus and clay particles have been moved downwards (eluviated) by water
>
> Ea: generally 'bleached' through removal of organic and sesquioxide coatings to particles; lacks mottles
> Eb: uniformly disseminated iron oxides; brownish
> Eg: partially gleyed
>
> *B horizons*: mineral horizons into which material has been moved by water (illuviated), generally from above
>
> Bf: iron pan; often enriched with organic matter and aluminium, although this may not be evident in the field (f = Fe, iron)
> Bg: partially gleyed, often with blackish manganese oxide (pyrolusite) mottles
> Bh: concentration of organic matter, with some iron and aluminium oxides
> Bs: concentration of iron and/or aluminium sesquioxides; typically rusty red/orange
> Bt: concentration of clay (German = *Ton*), often as coatings on ped (crumb particle) surfaces
> Bw: colour and/or structural development, but no accumulation of other materials
>
> *C horizons*: unweathered parent material
>
> Ck: coatings of secondary calcium carbonate (<1%)
> Cy: secondary calcium sulphate, usually as crystals (gypsum)
> Cm: continuously cemented other than as iron pan
> Cr: weakly consolidated; roots only penetrate cracks
> Cu: unconsolidated, uncemented, ungleyed

surfaces to attack; and the mechanical effects of roots and animals, a part of bioturbation (see pp. 86–7). Chemical weathering includes hydration and hydrolysis, the action of water; acidolysis, the action of weak acids deriving from lightning (sulphuric and nitric acids) or carbon dioxide (carbonic acid); and oxidation, often expressed as rusty colours of iron oxides. Biological

Soil horizons and soil types

Soil types

Here we list some of the more widespread European soils. US equivalents are given where there is matching of the two systems:

(A)C: raw mineral soils with a superficial organic horizon, usually young. US, entisol.

AC: lithomorphic soils, little-differentiated, containing organic matter, usually well-drained; shallow calcareous rendsinas and non-calcareous rankers, deep chernozems. US, mollisol.

ABC: brown soils, with a weathered Bw horizon; brown alluvial soils, brown calcareous earths and brownearths. Cuts across several US categories.

AEBtC: argillic brownearths, brown soils with a Bt horizon. US, alfisol.

AEBh/BsC: Podsols, showing strong downward movement of materials, leached from an E horizon and redeposited in Bh, Bs and sometimes Bt horizons. Brown podsolic soils and podsols, well-drained. US, spodosols.

Gleys: strongly affected by partial or total waterlogging. Surface-water gleys have impeded drainage, Eg horizons, within the soil; they include gley-podsols and stagnopodsols. Groundwater gleys have impeded drainage, Bg horizons, at depth; they include stagnogleys. US, in several orders.

Peat soils: with a deep O horizon. US, histosols.

Other US orders are: vertisols with high clay content which shrinks and swells; inceptisols, slightly developed but more so than entisols; aridisols of arid regions; ultisols, with argillic horizon, strongly weathered and typically in older landscapes; and oxisols, strongly weathered with little other than quartz and metal oxides (formerly laterites).

weathering is mainly the action of organic acids, deriving from plants, animals, fungi and bacteria.

After initial weathering, physical processes become subordinate to biological and chemical ones as plants colonise the surface. *Humus* formation sets in, and for this, plant tissues such as leaf litter are a major source. Decay may be underway before the leaf reaches the soil surface, whereupon bacteria, actinomycetes and fungi become active in breaking down the tissue and larger organic molecules. The decay pathway of incoming litter is influenced by the soil biota, which itself reflects the nature and decay of the organic component. In fresh leaf litter, the ratio of measurable carbon to nitrogen (the C:N ratio) is typically high: values of 30 to 50 are common. As the litter decomposes to soil humus, the C:N ratio falls to a value characteristic of the soil type, often in the range 10 to 20.

Of particular importance are the clay minerals which combine with decaying organic matter to form *clay-humus complexes*. These are vital in the retention of ions and nutrients in the soil, and in supplying minerals which are released into the soil water, providing essential ingredients for plant growth. Biological activity of mites, earthworms and enchytraeid worms in reworking the clay-humus complex forms the soil crumb structure in which soil material is intimately associated with water-filled spaces and in which the main chemical and physical reactions and biological activities take place. The crumb structure is maintained by cations, especially calcium, and organic matter. It is a fundamental part of soil fertility, just as much as is chemical and organic composition, for without it, the soil is reduced to a structureless mass.

Soil processes and soil horizons

Processes within the soil that are fundamental to its survival are the cycling of water and nutrients. Plants bring up minerals from the subsoil and return them to the soil in leaf-fall and at death. Animals return nutrients to the soil as dung and in their corpses. Worms are important in the incorporation of surface litter, passing soil and organic matter through their guts and ejecting it, partially digested, as casts. In some soils, moles are important, and in chernozems and other prairie blackearths, burrowing rodents make a major contribution to the cycling of organic matter. Water is brought up from the soil or subsoil, released into the atmosphere as a product of photosynthesis and as a cooling mechanism (transpiration), and returned to the soil as precipitation and condensation.

Processes in the soil form *horizons*, seen in section as roughly horizontal bands of different colours, textures and consistencies (Box 3.4). They are formed by the downward or upward movement and deposition of solutes or materials or by *in situ* change (Fig. 3.4). They are not stratigraphic layers. Some of the main processes involved are: mineralisation, the reduction and fixing of organic matter; humification, the combination of organic and inorganic matter as humus; oxidation, full aeration, giving rust-red colours; reduction, reduced or absent aeration, giving orange-grey mottling or grey colours respectively; leaching and eluviation, removal of materials from upper, A and E, to lower, B, horizons, often under conditions of high rainfall, free drainage and low pH; podsolisation, removal of iron oxides and humus from E horizons and their deposition as Bf and Bh horizons lower down, often under similar conditions to leaching and eluviation; lessivation, removal of clay from E and its deposition as a Bt horizon; and gleying, partial or total waterlogging, giving orange-grey-mottled or grey horizons, usually in conditions of impeded drainage and high water-tables (Box 3.4). In semi-arid and tropical conditions, removal of silica leads to laterisation, while salts like calcite and gypsum can be moved up or down the profile, leading to the formation of crusts by processes of calcification and salinisation.

Soil profiles and soil types

The sequence of horizons from surface to unweathered substratum is the *soil profile*. Classification is based on the presence and absence of horizons, and a given soil type may show few or many (Box 3.4). A cultivated soil on calcareous loess may show only Ap and Ck horizons, whereas a leached soil under coniferous woodland may have Of, Oh, Ah, Ea, Bh, Bs and Cr horizons. In classifying soils, UK (Avery 1973; Burnham 1980) and European (FAO 1974) schemes combine character, climate and process; before classification, some interpretation is needed. In the USA, in order to avoid this subjectivity, soils are placed in a hierarchical classification based on the presence or absence of characteristic horizons (Soil Survey Staff 1975; 1990).

Environmental groupings

Globally, soil types are related to climate. Organic matter, water movement and the concomitant translocation of soluble chemicals, clay particles and humus are important factors (Fig. 3.4). In temperate climates, where precipitation exceeds evaporation and transpiration, there is net downward movement, with horizons of translocated iron oxides, humus and clay (alfisols, spodosols). Organic matter content is often high. In the wet tropics, leaching and chemical weathering lead to rapid loss of organic matter and silica, and the development of thick horizons of aluminium oxides (oxisols). Tropical woodland soils have little organic matter, most of the biomass being in the vegetation. In arid regimes, where evapotranspiration exceeds precipitation, net movement is upwards, leading to the crystallisation of gypsum, calcite or other salts in the upper parts of the profile (aridisols) or as a crust at the surface. In continental interiors, where there are warm dry summers and cold winters, thick organic-rich humic layers overlie the subsoil directly, as with the Russian steppe blackearths or chernozems and the North American

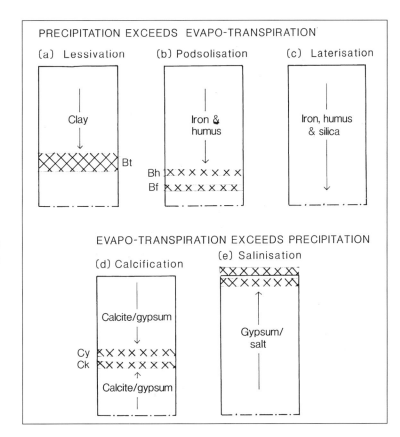

Figure 3.4 Translocation of materials and chemicals in soil profiles; for horizons, see Box 3.4. (a) Argillic brownearth or alfisol; (b) podsol or spodosol; (c) oxisol or laterite, enhanced in aluminium sesquioxides; (d) aridisol with a balance of leaching and capillary rise; (e) aridisol, with surface crust.

prairie soils (mollisols). In soils of low biological activity, as under coniferous woodland where the litter is acidic, or in cool climates, decomposition of organic matter is so slow that it occurs as raw humus, sometimes as recognisable plant remains. Leaching is prevalent, giving rise to spodic, Bh and Bf, horizons (spodosols). It is often those plants which can tolerate the poorer soils that produce the most acidic litter, so acidification and nutrient loss are enhanced. On this scale, soils are referred to as classes (UK and European schemes) or orders (US scheme). They are related to climate (and vegetation) zones and are referred to as zonal soils.

Intra-zonally, climate is less important. Soils are related more to hydrology, geology and topography. One factor is cultivation, which maintains soils in a youthful state through continuous contact with the substratum. Another is the removal of vegetation, especially trees. Poor land management, as with over-cropping or lack of manuring, is especially deleterious. These processes slow the cycling of water, nutrients and organic matter, reduce the organic and nutrient content of the soil and its biological activity, and weaken the crumb structure. There is often a decrease in pH which enhances these trends. Free soil particles clog the pores of the crumb structure, often with fine charcoal from human or natural fires. In extreme cases the soil breaks down to mineral and organic particles and loses its structure entirely. Processes which had been operating in equilibrium start to intensify or become unidirectional.

Taking the brown forest soil as the zonal type of temperate climate, the following tracks are

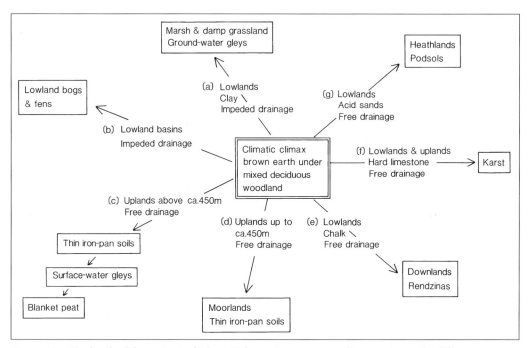

Figure 3.5 Tracks of soil formation in the later Holocene in a temperate climate region under different conditions of topography, subsoil and drainage; for letters, see text (p. 35).

suggested (Fig. 3.5): (a) On lowland clays of impeded drainage, there is partial waterlogging or gleying, with orange-grey-mottled or grey groundwater gleys. (b) The formation of valley bogs and fens takes place in really wet areas, where there may have been no previous soil at all. (c) In uplands, a combination of high precipitation, low temperatures and low biological activity lead to acidification and leaching (Fig. 3.4b) which result in the formation of an iron pan; this in turn, because it impedes drainage, leads to surface-water stagnopodsols and, ultimately, blanket peat. (d) In uplands at lower altitudes, thin iron-pan soils without a covering of blanket peat are widespread. (e) On chalk and other soft limestones, thin humus rendsinas form with no B horizon; these erode under cultivation forming *colluvium* in valley bottoms and *lynchets* on valley sides. (f) On hard limestones, there may be total erosion to bare rock pavement, as in the Irish Burren and other karst areas. (g) On well-drained acidic sandstones in lowland areas there is leaching with the formation of iron/humus podsols; break-up of the surface vegetation leads to gully and wind erosion. Lack of agricultural activity in these areas may have led to their impoverishment (Limbrey 1975). On neutral well-drained loams, downward movement of clay, known as *lessivation*, leads to the formation of argillic brownearths, also called *sols lessivés* (Duchaufour 1982) (Fig. 3.6).

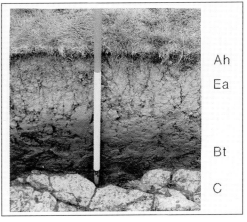

Figure 3.6 Argillic brownearth profile on Carboniferous Limestone, Malham, Yorkshire; the Bt horizon is dark because its clay content has caused it to dry out more slowly than the rest of the profile. In fact, this profile is complex since the lower part is a cultivated palaeosol; it contains charcoal flecks, and magnetic susceptibility values are relatively high (p. 115).

This gives a genetic perspective to soil diversity at the local and regional scale, although some of the processes such as lessivation and podsolisation can take place in zonal soils and under temperate woodland outside human influence (Fisher 1982). The soils are termed 'azonal' because they are immature or in a state of flux and not fully adjusted to climate. They can be equated with the disequilibrium communities of biology (p. 54). Climate, however, still provides the limits to the formation of particular types; for example, in temperate areas, soils are based on a silica structure because the temperature range does not allow its dissolution.

Other groupings

Transcending such classifications are groupings such as susceptibility to drought, erosion potential and fertility/productivity, which can be of greater use when we want to understand past human activities in relation to soil types and their potential for cultivation. Soil maps (see p. 108; Fig. 9.2) show these properties, first as associations – 'groups of topographically related soils developed on the same parent material' (Davidson 1980, 25) – and then as smaller units, the soil series, often named after a type locality.

Complications

What happens when we dig a soil pit? Rarely do we find a zonal profile or even one of our regional types. Frequently, unexpected profiles are revealed, with layers of charcoal or human occupation, old cultivation horizons, truncations or sediment. Younger soils may overlie the truncated profiles of older ones, the older profiles providing the subsoil for the younger ones. Relic soils occur in unglaciated and level areas which have not been eroded, such as upland plateaux (Fig. 3.6). The lower profile, often with a prominent Bt horizon, may be of considerable age, going back into Pleistocene or earlier times (ultisols). This provides the C horizon for later soils. In many upland areas in temperate Europe, often just beyond the altitudinal or pH limits of present-day agriculture, a podsol profile with a thin iron pan (Bf horizon) lies above an older, unleached one with a Bt or (B) horizon (Fig. 3.7). There is often evidence of burning and cultivation at the interface of the two profiles. Soils of this complexity are referred to as *compound or multiple profiles*.

In fact, it is probably hard to find any soil that is totally 'natural', and there are different degrees of human influence. On slopes, cultivation leads to the formation of terraces or lynchets, either by erosion and build-up against a boundary or deliberately, as in areas of

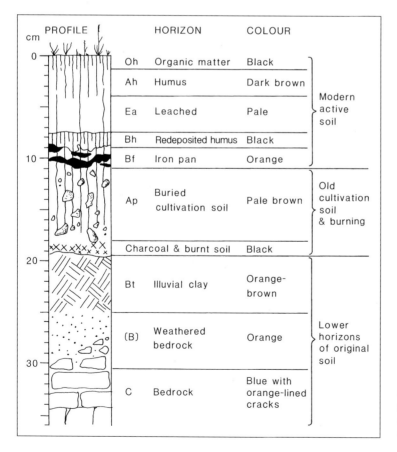

Figure 3.7 Complex soil profile, typical of British uplands around the limits of cultivation (pp. 157–8).

steep terrain like the Andes and semi-arid climates like the eastern Mediterranean countries (pp. 209–10). Some soils are *in situ*, modified by cultivation and manuring. Deep ploughing brings up subsoil material and may lead to thickening of the A horizon. Keeping animals at night on cultivated soils, and the manuring of such soils from settlement refuse and night-soil can lead to a great thickening of the A horizon. Plaggen soils (plaggepts) in the Netherlands are formed by depositing soiled animal bedding, such as heather and turf, to increase fertility and to raise the land surface (Groenman-van Waateringe and Robinson 1988). All these processes increase local contrasts in land productivity, with depleted or eroded areas becoming poorer, while upgraded areas become richer.

Summary of groupings

Different reasons for classification have led to different schemes. Avery (1990, 37–80) gives a good summary of the major classifications in use around the world, and their approximate equivalence.

1. Global schemes of the Soil Survey of England and Wales (Avery 1973; Burnham 1980), USA (Soil Survey Staff 1975; 1990), and the FAO (1974) with their major groupings of zonal soils. These use names like podsolic soils (UK), spodosols (USA), and arenosols (FAO), sometimes called 'great soil groups'. They are used when we are studying long-term changes in relation to climate.

2. UK scheme of soil types, with names like brownearths, argillic brownearths and stagnopodsols. These encompass climate, processes and characters simultaneously (Duchaufour 1982, 161), and are useful if we are studying processes and change at the intra-zonal scale, as in relation to change within the Holocene.

3. UK (and other) schemes of 'soil series' (Fig. 9.2) which name soils at a local scale and which are useful names when mapping a relatively small area.

4. Within profiles, horizons are named according to the processes, like Eh and Bs, and character, for example, non-calcareous sandy loam.

5. *Fossil soils*, if well characterised in terms of profile, distribution and age, are named after type localities (Box 10.1).

6. In archaeology, where we relate soils to environment or land-use, it is common to refer to soils by environment (floodplain, Alpine or anthropogenic), character (sandy or calcareous), process (leached, gleyed or young), or suitability for particular uses (fertile, good or poor) (FAO 1974).

SEDIMENTATION

General properties

Sedimentation is the process by which material builds up at the surface of the earth. It reflects disruption of the cycling of nutrients and water through the soil. Sedimentation usually occurs in one of four ways: where surface instability leads to the release of rock particles, their transportation and deposition; where dead plant remains do not decompose but build up to form peat; where plants and animals secrete mineral matter as their skeleton or excretory matter, usually calcium carbonate, but sometimes silica (diatom skeletons) or phosphate (bird guano); and human constructions.

Characteristics of sedimentation

Sedimentation slows the growth of rooted plants, although many, like sand-dune, lake-bottom and river-bed plants are adapted to it. It results in the upward accumulation of material, although the build-up can be so slow that contemporaneous soil formation occurs. Sediment characteristics are established as the sediments are laid down, although consolidation by water percolation and mineral precipitation can occur later (Table 3.2). Sedimentation lays down material which is dead or quickly becomes dead, and which does not support the cycling of nutrients and energy. Some sediments are formed by biological processes, either as a part of the life of the organisms themselves, as with coral and algal reefs, or by their accumulations at death, as with peat. Additionally, the upper layers of most sediments have their own biota, but these need not be responsible for the formation of the sediment (p. 40). An important characteristic is that sedimentation may be local, occurring in basins like caves and lakes. Where processes are not linked to basins, as with wind deposition and glaciation, sedimentation is more widespread. There is also great variation in thickness, with sediments several kilometres thick, in the case of Pliocene river and lake sediments in East Africa, or so thin as to be invisible to the naked eye, as with traces of volcanic ash (tephra) in a peat bog. Sedimentation in temperate climate terrestrial

Table 3.2 *Classification of clastic sedimentation processes and the sediments*

Process	**Sediment type**
High-energy transport	
Ice	Till or boulder clay (=diamicton)
Water	Rounded, bedded gravel
Gravity	Angular scree
Sub-aerial	Coombe rock and rock streams
Moderate-energy transport	
Water	Rounded, bedded sand
Wind	Subangular to rounded, bedded sand
Sub-aerial	Loamy, stony colluvium weakly bedded, poorly sorted
Low-energy transport	
Water	Silts and clays
Wind	Silt (=loess)

Note: Gravel consolidates to conglomerate, scree and coombe rock to breccia contexts is always intermittent, reflecting instability and change. For most of the timespan and in most areas of human occupation, sediments were not and are not forming. It is only in arable land that we see extensive areas of broken surfaces and even here erosion is episodic. Even apparently dynamic situations are stable enough to be colonised by plants and animals. In a flowing stream, there will be vegetation growing on the stones, and in the surface of active coastal storm beaches there is abundant arthropod life. In arid and semi-arid areas there is constant reworking of the ground surface, but even here sedimentation is episodic. Finally, sedimentation alters topography: tufa dams, algal and coral reefs, blankets of loess, deposition in caves and human constructions like cities affect altitude, conformation of the land surface, hydrology and climate.

Classification of sedimentation processes

Sedimentation involving transportation gives rise to *allogenic sediments*, generally made up of mineral particles or clasts, but which may be composed of biological materials, such as drift wood (Fig. 3.8). There are three main stages in their formation: weathering, transportation and deposition. Weathering (pp. 29–31) gives rise to the materials of sedimentation. Transportation takes place at various rates (Table 3.2) and during this process erosion occurs, again at various scales, from the massive scouring of valleys and lands by glaciers, through the small striae left by the ice on rock surfaces or the marks of ploughs, to that of the particles themselves. Deposition depends on the presence of a basin, or some other place where transportation is slowed or halted, and the products of much erosion are washed out into the sea. Rate and distance of transport and consistency of process are important determinants of sediment type, reflected in clast size, shape and sorting (Table 3.3). Scree, or rockfall formed by gravity, is made up of large, angular and uneroded particles. In contrast, a gravel or sand that has seen a long history of erosion and is far from its origin is made up of rounded particles. Where there is a single transport and deposition mechanism, well-sorted sediments, i.e. those having a narrow size

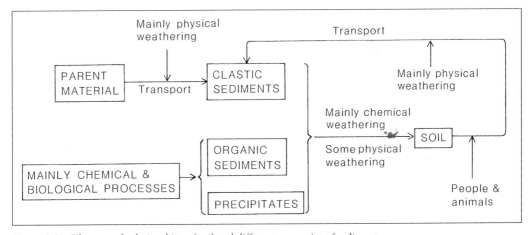

Figure 3.8 The general relationships of soil and different categories of sediment.

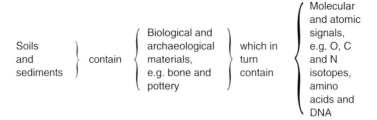

Figure 3.9 Relationship between conventional soils and sediments and the archaeological and biological remains within them (cf. Fig. 12.1).

distribution, may be formed, such as gravel in flowing water, silt (or loess) by wind, and clay in still water (Fig. 4.2). A mixture of processes, especially prominent in terrestrial (sub-aerial) environments, like rilling and gullying, creates ill-sorted loam.

Sediments formed without transportation are *autogenic*. They are usually of biological origin, or *biogenic*, and are subdivided into crystalline sediments formed by precipitation such as stalagmite and coral, and those formed from organic materials like peat and coal. Precipitation generally occurs in the skeletons of organisms, and often, especially in aquatic environments, these are anchored to the substratum. They are not destroyed at the death of the organism because they are built over and buried by the skeletons of the next generation. Furthermore, the biological materials themselves preserve their own sedimentary history (Fig. 3.9), whether in the form of physical features like ring or layer thickness, or of stable isotopes (p. 148), or DNA and other biomolecules.

All sediments preserve several episodes of history. Thus with biogenic sediments, some organisms may be derived from earlier sediments, some lived in the materials which are deposited, as with soil biota, some, like corals and sand-worms, are responsible for the sedimentation process itself, while yet others live in the sediments after they have formed. Some organisms support epiphytic and epizoic communities (Fig. 3.10).

Equally with clastic sediments, the particles can have different origins and histories before weathering even begins, some having been transported hundreds of kilometres by wind or ice, others being of local origin. During transportation different degrees of clast erosion take place, while other processes result in particle orientation and structures as the sediment forms. Finally diagenic processes such as cementation and soil formation affect the sediment after it has formed (Fig. 3.11).

Figure 3.10 Oyster shell outer surface, providing a substratum for boring worms, marine snails and bivalves.

ECOLOGY AND ENVIRONMENT

The onset, continuation and cessation of sedimentation

How is sedimentation controlled? The onset of sedimentation is brought about by environmental change. This may be a single factor like a sea-level rise, or it may be related to a complex of factors but require a trigger of just one factor, perhaps just a single storm. Soil erosion, for example, may be brought about by an increase in precipitation but this may not cause erosion unless the soil has been primed by bad land-use weakening the crumb structure in the first place. Sedimentation continues because the conditions created by its initiation have a positive feedback. Erosion leads to bare soil and rock, providing further materials for sedimentation. In some cases of precipitation, as with algal reefs, the development of the sediment provides more suitable conditions for succeeding colonisation so sedimentation is enhanced. Blanket peat changes the environment and permits some species of plant and animal to flourish, especially those which increase the acidity of the environment and thus reduce its biological decay. The cessation of sedimentation may be brought about by a further, extrinsic, environmental change or by the sediments reaching a height where they can no longer accumulate.

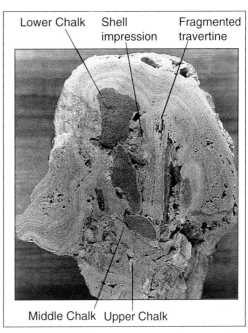

Figure 3.11   Section of travertine from Marsworth, Hertfordshire, encasing chalk fragments of different origins and shattered travertine from an earlier period of formation; note the secondary travertine formation around the chalk clasts.

Sediment classification and terminology

A classification of sediments based on their properties (as opposed to processes, cf. Table 3.2) is shown in Table 3.3. This is a hybrid classification in that some process and interpretative, rather than character, names are used – like scree, boulder clay and tufa – but these are a convenient short-hand for what would otherwise be quite long descriptions.

Sediment units

Sediment, or *lithostratigraphical*, units are classified on a hierarchical basis, and it is important to follow international rules (Bowen 1978). The **group**, as the highest level, includes a wide variety of deposits, such as all the peats, lacustrine and fluvio-lacustrine units in a region attributed to one interglacial. The **formation** is widespread, covering several square kilometres, and should be mappable as a tabular body of strata, e.g. blanket

Table 3.3 *Sediment classification*

Composition	Environment
1 CLASTIC SEDIMENTS	
1a Moderately to well sorted	
Gravel, >2.0 mm, rounded; consolidated as conglomerate	Glacial outwash, sea storm beach, fast-flowing river
Scree, >2.0 mm, angular; consolidated as breccia	Cold climates, primary ditch fills
Sand, 2.0–0.06/0.05 mm; consolidated as sandrock	Aeolian = coversand, water-lain as sea beach or river deposits
Silt, 0.06/0.05–0.002 mm (= 60–2 µm); consolidated as siltstone	Usually aeolian = loess
Clay, <2.0 µm; consolidated as claystone	Still and/or deep water, e.g. lakes and oceans
1b Poorly sorted	
Till or boulder clay	Ice-sheet or glacier
Solifluxion/gelifluxion	Cold-climate slope deposits
Stony, humic loam	Slopewash, e.g. under cultivation by rill erosion (= colluvium)
1c Volcanic deposits, e.g. tephra, acids, pyroclasts, lava	Various volcanic environments
2 ORGANIC SEDIMENTS	
2a Peat	
Cladium	Fen
Sphagnum, Eriophorum, Calluna	Raised bog
Phragmites	Reedswamp
Wood and tree stumps	Carr
2b Lake mud or gyttja	Lake bottom
3 PRECIPITATES	
3a Calcareous	
Tufa	Springs, rivers, lakes, caves; deposited by blue-green algae (Fig. 3.11)
Travertine/stalagmite is a hardened, crystalline form of Tufa	
Lake marl	Lake bottoms, formed by Characean and pelagic algae
3b Bog iron, sulphates, sodas, silicates, phosphates	Miscellaneous, mostly aquatic

peat or alluvium; this is the primary unit in lithostratigraphy. A **member** is a distinctive horizon within a formation which again is mappable, e.g. a volcanic ash layer within peat. **Beds** are usually made up of a single type of deposit – moss peat or calcite loam – in a member or formation; they are usually of local significance. In his work at Olduvai Gorge, Hay (1976) used a different system. The bed was the fundamental unit, below which were lithofacies: sedimentary units of differing content and texture which were deposited at broadly the same time in a closely related series of environments. Archaeological deposits are sometimes amenable to classification by lithofacies (e.g. Gilbertson 1995). Classifications can be subjective and change with the accumulation of data. But whatever the scheme, use of a named unit must refer to a type locality and section, and in the case of a formation, a tabular, mappable unit.

Spatial distribution and sequences

Spatially, sedimentation is patchy and graded. Patchiness can be zonal, relating to factors over large areas, like climate; it may be regional or local, relating to vegetation cover and human land-use; and it may relate to the sediment basin itself. Gradation refers to the decreasing energy of deposition and decreasing particle size of sediment as one moves from away from the sediment source (Figs 2.1 and 4.2). In glacial and periglacial climates physical weathering predominates and sediments are usually clastic. In the Pleistocene, within each cold stage, there is a pattern of change which is spatial and temporal. Frost-heaving (*cryoturbation*), whereby one sediment layer is involuted into another, is common. Ground-ice masses, occurring as crack infillings (ice-wedges), lenses or mounds (pingos) and continuous layers, are widespread in the coldest periglacial regions. On the melting of the ice during climatic warming, these features are infilled with sediment, often different to that of the surrounding material. Cryoturbation features and ground-ice casts often leave their mark as 'patterned ground', so-called because of the patterns they produce as seen from the air. In warm temperate climates, organic sediments and precipitates predominate; erosion and the formation of allogenic sediments are minimal except on coasts and in river valleys. Biological activity is higher than in cooler or drier climates and results in a variety of localised features like burrows and *tree-throw pits*. There is a sequence of algal precipitates in the Late-glacial and Holocene, with lake marls being common early on, while tufa, formed in swamps, streams and caves, is more a feature of the middle stages. So too with peaty deposits: nutrient-rich fen peat tends to occur in the earlier stages of the Holocene, while blanket peat is more a feature of the middle to later stages. In arid and semi-arid climates where vegetation is sparse and land surfaces poorly protected by vegetation and soils, erosion is common and minerogenic sediments, particularly alluvium and sub-aerial deposits, widespread. Long periods of dryness followed by rainstorms enhance these processes. Aeolian processes lead to widespread deposits of loess and coversand. Precipitates of calcium carbonate, gypsum and chlorides formed by evaporation within and at the soil surface (aridisols, Fig. 3.4d) and in playa lakes are common. At regional and local scales, topography gives rise to specific sediment types, especially with rivers and coasts where sedimentation can cut across the zonal pattern. Again, gradation is a feature, especially in rivers in which particle size decreases as one moves from source to mouth. Areas of local sediment sumps, such as karst areas with their caves, recently deglaciated areas with their lakes and wetlands, and areas of intensive human management with field and settlement ditches, create sediment patchiness at the regional and local scales.

A word of warning

Sedimentation is referred to variously in terms of environment (e.g. cold-climate), character (e.g. high-energy) and process (e.g. fluviatile). Care must be taken because these terms mask a range of processes and characters. 'Ocean sedimentation', for example, includes clastic and biogenic deposition and precipitation, and can also refer to contributions from ocean waters (pelagic) and continents (terrigenous) (Bowen 1978). *Colluviation* refers to slope deposition which may be of cold-climate or warm-climate origin. It is a term which is often used as short-hand for deposition through agricultural activity, but Allen (1992) has shown that there is a range of processes, even for this meaning. *Alluviation* is often taken to mean the deposition of fine river-lain materials by overbank flooding; but it can equally apply to the precipitation of calcite loam, deposition in channels and backswamps and of Pleistocene gravels (pp. 160–4).

Four

Ecosystems

The last chapter reviewed some components of environments and established some important terminology. If we are to understand modern environments, however, we need more than a list of characters. The challenge which we now take up is to understand the dynamics of the ecosystems from which the lists are derived.

Definition and Characteristics of Ecosystems

Interactions of biotic communities with the physical environment within a defined area constitute an ecosystem. In reality, there is probably no such thing as an independent ecosystem. A lake is a good example: it has a clear boundary, but even in a lake without inflowing or outflowing rivers, there are links to the outside. There is input from land erosion, minerals from the atmosphere, and plants and animals by active and passive dispersal, and there is loss by seepage, evaporation, and emigration of plants and animals. Ecosystems differ from environments and habitats in being dynamic, and they differ from niches in referring to multi-species communities. They are characterised by their spatial extent, boundary clarity, physical structure, biological populations and energy flow.

Ecosystem size is, to a degree, defined by the observer, but should be sufficient to encompass interactions of geology, soils, biota and atmosphere. The crevices of a boulder in the Arctic tundra will meet these criteria, as would the entire planet. When it comes down to human interaction, woods, prairies, rivers, valleys and lakes are what we usually have in mind. Boundary clarity is important in the definition of ecosystems – as with lakes – but it also has an importance in its own right, namely in the identification of ecotones (see p. 53). The physical structure within which biological populations cycle matter and utilise energy includes topography (Box 3.2), vegetation layering and architecture, soil and sediment structures like raised bogs and reefs, and animal structures like burrows and termite mounds. 'Layering' here refers to the different levels of vegetation such as mycorrhizal fungi, leaf litter at the soil surface, and the herb, shrub and tree layers, which also structure the distribution of other plants and animals.

The biotic community of an ecosystem characterises its complexity in terms of competitive and energetic relationships (pp. 26–8). Ecosystems embody energetics: the passage of nutrients and energy within the physical and biological environments, through food webs within the biological communities, and by inputs and losses from these webs. The rate of primary production, the number of trophic levels, and the location of different categories of biomass are key components. Primary productivity in land ecosystems varies with latitude and altitude, with the most productive ecosystems being in the wet tropics and at lower altitudes, most especially flood plains and deltas. In the oceans, productivity tends to be high in coastal waters, or where upwelling or turbulence brings nutrients towards the surface, and is low over vast tracts of ocean (e.g. Mielke 1989, 126–9).

Space

Energy is variably distributed in ecosystems, sometimes being concentrated, sometimes dispersed. Plant and animal life-cycles, feeding strategies and general mobility create spatio-temporal variations in the aggregation and dispersal of populations (Box 3.2). Human technologies, food-getting strategies and settlement systems often respond to these aggregations. Raw materials, like timber, stone, clay, metal ores, seaweed and guano, also often occur in concentrations, allowing ease of collection and extraction, and thus cutting down on the time and energy involved. Equally, concentrations of particular kinds of topographical feature – pits, lakes, caves, coastlines – also allow energy-saving in transport, the location of food and the construction of dwellings.

All artefacts, biological materials, soils, sediments and lands reflect or relate to a wider area than themselves. Put simply, species diversity depends on the area of habitat, though this apparently simple species–area relationship is difficult to disentangle from higher species diversity resulting from the greater structural diversity of larger areas. An ecosystem is also constantly undergoing exchanges with environments beyond its boundary. Rainfall brings in water, dust and dissolved compounds; plants and animals come and go both actively and passively; and humans may deliberately or inadvertently move materials and species in or out of the system.

Certain ecosystems, often very localised, can develop particular aspects of the wider environment and processes which may not otherwise be so apparent. We see this in small localities which are the focus of human activities of far wider, but otherwise cryptic, social relevance. The use of rubbish (Hodder 1982a), the different areas of Neolithic camps, the elaborate entrances of Iron Age hillforts, and the patterned emplacement in pits and chambered tombs of artefacts, economic material and human remains (Thomas and Whittle 1986) are all examples of this kind of development. Pressures and constraints from outside the ecosystem may have distinctive consequences in speciation and behaviour. A species can be segmented into geographical populations of different ecologies in response to an environmental gradient (Fig. 3.1), or the same economic, climatic or social pressures on a wide scale can lead to locally different land-use according to conditions of soil and topography. At a smaller scale, topography develops the same overall influences in different ways, as in variations in vegetation and soils along a catena (see Fig. 4.5) or in sediment sumps like lakes where muds are deposited in the abyssal zone and tufa banks or reefs around the fringes.

Time

Embedded in every ecosystem are four levels of time (Fig. 4.1). (a) Potential for change. Some ecosystems are fragile (or brittle); they are easily changed and replaced by other ecosystems, as with some kinds of coniferous woodland. Others are resilient, and even when perturbed are

ECOSYSTEMS

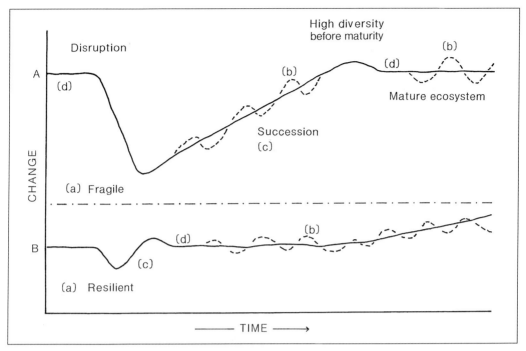

Figure 4.1 Four levels of time: (a) potential for change seen in the contrast between resilient, A, and fragile, B, ecosystems; (b) short-term, cyclical; (c) succession; (d) long-term. The horizontal parts of the lines represent ecosystem maturity, the sloping parts degrees of variation from this (pp. 46–7).

likely to recover readily, as with rocky seashores. (b) Short-term changes. All ecosystems change on a scale of a few years. This may be readjustment or succession (p. 51) after environmental change, inherent change in an ecotone, recurrent rejuvenation as with annual flooding, or cyclical variations in species' abundances in relation to predator–prey interactions. (c) Age. This includes the successional stages through which the ecosystem went and its duration since the last major environmental disruption. (d) Deeper history. Continental movement, archipelago formation, vulcanicity, uplift and sedimentation have created a diversity of time base-levels. Even in relatively recent origins there is an embeddedness of geology, fossil topography, adaptiveness to past environmental stasis or change, and genetic resilience to inbreeding (Table 3.1), all of which can show through into contemporary ecosystems.

GRADIENTS

One way of grasping the diversity of ecosystems is to envisage a series of gradients, such as the gradient from coarser to finer sediments between continents and oceans (Fig. 2.1). The largest gradients are climatic (Fig. 2.2). Blocks of varying primary production, biomass distributions, biota, soils and sediments and their different land properties of weathering and change are arranged in an essentially latitudinal zonation: ice-caps, tundra and taiga (with attendant permafrost, itself a legacy of the Pleistocene), temperate woodland,

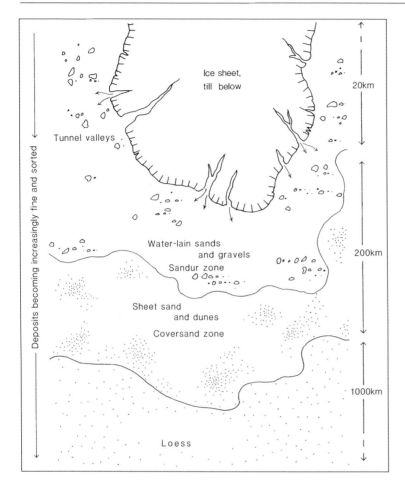

Figure 4.2 Variation of sediments and land at the sub-continental scale in relation to an ice-sheet (cf. Fig. 2.1).

Mediterranean vegetation, semi-arid, desert and tropical rain forest. There is also some longitudinal zonation – temperate woodland in oceanic areas, savanna and other grassland in the continental interiors, and deserts along some continental edges. Superimposed on all of this is zonation by altitude, giving gradients typically from woodland to grassland to bare rock. In glaciated areas, there is decreasing energy of deposition and increasing fineness and sorting of materials away from the ice (Fig. 4.2). The ice itself erodes and smooths rock surfaces, carves U-cross-sectioned valleys in uplands, and deposits ill-sorted till; the uneven till and rock surface, sometimes from erosion, sometimes from pockets of remnant ice, has created a land of lakes which today extends across much of Canada and picks up again in Ireland across into northern Germany, Poland and around into Finland and western Russia. Beyond the ice, there are water-transported outwash sands and gravels from beneath the ice-sheets and glaciers; these are called *sandurs* after the Norse term for this type of terrain in Iceland today. Two zones of successively finer wind-lain deposits – *coversands* (sand-size particles), terrain referred to as *Geest* in the Low Countries and north-western Germany, and *loess* (silt-size particles) – lie beyond this, each

ECOSYSTEMS

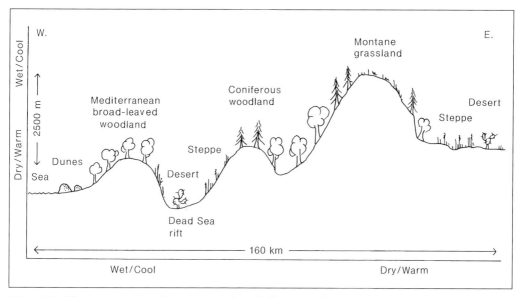

Figure 4.3   West–east transect at the eastern end of the Mediterranean showing local variation of vegetation as related to altitude and regional variation as related to distance from the coast, in a generally semi-arid climate zone.

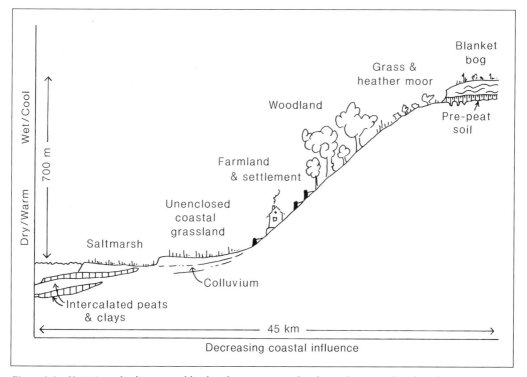

Figure 4.4   Variation of sediments and land at the intra-regional scale in relation to altitude and coastal influence in a temperate maritime region modified by humans (cf. Fig. 4.6 and Table 7.2).

giving a different substratum to the soils over extensive areas from that of the native rock. Furthest away from the ice-sheet, clays are deposited in estuaries and seas.

There are shorter gradients from high mountains, through lowlands, to the coast, with climate as related to altitude and coastal influence being the main factor. In semi-arid climates as in the southern and eastern Mediterranean (Fig. 4.3), there is a general horizontal zonation from coastal woodlands through inland steppe to desert, superimposed on which are altitudinal zonations of piedmont woodlands of various types and montane grassland. In temperate climates (Fig. 4.4), mixed deciduous woodland is the climax vegetation, but above a certain height trees are unable to grow and ecosystems are of low shrub and herbaceous species, with soil instability and permanent snow and ice. At the coast, there is a strip of land which is open, except in sheltered areas; sometimes this may be extensive saltmarsh. These gradients are microcosms of global ecosystem zonations, but with the important difference from the human point of view that distances between the zones are shorter, allowing greater ease of access. This is shown particularly in coastal ecosystems (pp. 165–8). Highest are fossil cliffs and caves suitable for human occupation, with raised beaches and river terraces with sands and gravels suitable for cultivation, while lower down are plains of clay supporting saltmarsh and other wetland ecosystems (cf. Figs 12.12 and 12.13). Post-Glacial sea-level rise in some parts of the world flooded extensive coastal saltmarsh, so that maritime ecosystems were brought more within the sphere of inland ones, thus affording human communities greater choice of land types in a shorter spatial gradient.

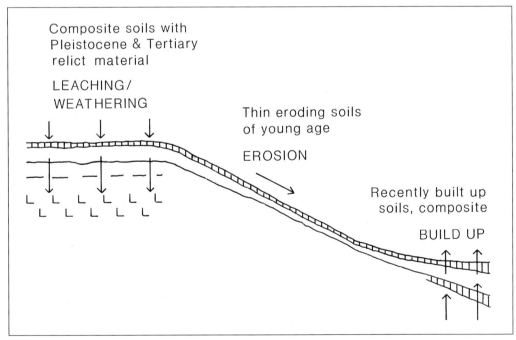

Figure 4.5 Variation of soil and land in a non-glaciated area at the local scale as related to slope and the absence of erosion on the plateau (cf. Fig. 3.5).

ECOSYSTEMS

Figure 4.6 Land variation at the local scale in a temperate climate region at the limits of cultivation; Gordale, near Malham, Yorkshire. This is essentially the upper part of an upland sequence (Fig. 4.4), although woodland is largely absent.

Locally, a useful concept is the *catena*. This was devised by soil scientists to describe changes in a soil profile along a slope, as the varying influences of leaching, downslope movement and build-up act at different points along the catena (Figs. 4.5 and 4.6). The concept can be applied to the response of the environment along any kind of gradient, such as salinity, wetness (Fig. 4.7), energy transfer (like the length of the growing season), and intensity of human activity. In some cases, catenas are related to a gradient in the timing of the last exposure of the substratum to physical weathering, thus giving rise to soils of different maturity. An example is the sequence of terraces along a river valley, where, because the higher terraces have usually formed earlier than the lower ones, the soils on them are increasingly immature as one moves down the valley side (Box 10.2).

SUCCESSION

Temporal changes go on at each point along a gradient. At their most complete, these see ecosystems developing from a lifeless to a climax state which is in equilibrium with the regional climate. This is succession and the complete sequence is termed a *sere* – a hydrosere starting from water, a xerosere from dry rock. The early stages see physical weathering, resulting in coarse minerogenic sediments (p. 39), with a specialised biota adapted to the rapid colonisation of immature lithosols. Because of surface instability, annual plants and other opportunist species are common. The later stages are often a product of gentler processes, resulting in finer clastic or autogenic sediments. Surface vegetation cover becomes continuous, soils are deeper and with a strong crumb structure, and plants and animals are typically equilibrium species (Box 4.1). Tracks of succession vary according to slope, climate, substratum and hydrology. In a valley profile (Fig. 4.5), surface stability and soil development may be rapid on the plateau, especially if there is relict soil material in place (pp. 127–8). On the slope, where bare rock may be exposed, erosion may delay the establishment of stable soils and vegetation. On the valley bottom, flooding or waterlogging may lead to a sequence of immature soils, sediments and wetland vegetation before deposits build up sufficiently to give rise to dryland. Ecosystem maturity is a combination of high species diversity, the establishment of numerous, often narrowly constrained, niches, and a length of time and period of unchanging conditions long enough to give rise to zonal soils and vegetation.

As originally presented, succession theory was largely based on observations of present-day spatial gradients without a knowledge of the temporal framework. It was an analogue approach; in archaeological terms a typology. With longer-term observations and more understanding of ecological change, we now know that things are more complex. Thus some seral stages do not progress to the next potential stage because colonisation by new species is inhibited. The ecosystem presents a state of inertia which allows it to continue even though new species are nearby to take over and are more suited to the prevailing environment. Indeed, zonal soils and vegetation may never be achieved if there is a preventative long-term factor like animal grazing or extreme topography. Only the stimuli of human interference, fire, disease or extreme weather may eventually cause a threshold to be reached so that succession continues (Smith 1965). We can also see the sequences

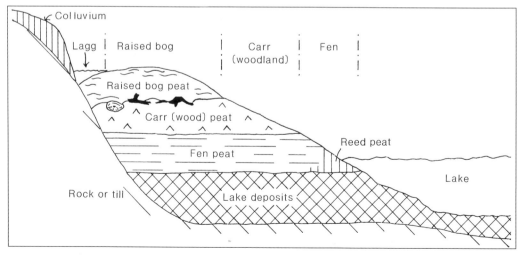

Figure 4.7 Relationship between vegetation and sediments at a lake edge, showing variation at the intra-locality scale as related specifically to a sediment sump.

themselves in the geological record, as in lake basins (Fig. 4.7). The catena is often from open water, through reedswamp, sedge fen, to woodland, suggesting the successional stages, but Walker (1970), studying actual temporal sequences, showed that the climax vegetation is not terrestrial woodland, although the succession passes through a woodland stage, but raised bog. Climate is overridden by topography and hydrology.

A sere in which an event intrudes and deflects the course of succession, like a climatic shift, short-term weather abnormality or fire, is known as a *plagiosere*. Once the factor is removed, recovery begins, although the stage in the sere at which this takes place and its track depend on the degree of perturbation (cf. abandonment, p. 76). Total soil erosion entails the succession beginning from bare rock and recovery is a long process. In temperate regions, damage by human woodland clearance and agriculture without complete soil removal entails some soil development and vegetational stages specific to the plagiosere, before the climax vegetation is reached. When there is a long period over which the ecosystem remains unchanged but with short-term pulses of change, the situation is known as *pulse stability* (Odum 1971). Examples include: burning which upgrades vegetation *vis à vis* the succulence of the shoots for herbivores even if reducing cover; flooding which enhances grassland in warming it early in the year, the basis of the water-meadows of post-medieval England; and alluviation, which, if incremental and not killing the vegetation, adds nutrients to the soil, as in the Nile valley.

HUMANS

The influences of humans since the end of the Ice Age are everywhere. In permanent farming and village societies, the most visible effect is the mosaic of ecosystems ranging from a few square metres to several tens of hectares, patchworks of different kinds of land for farming, village life and communications. Most vegetations are managed even though they may seem natural; fen, carr and reedswamp have management (which means time) connotations as well as being purely descriptive terms (Rackham 1986). In Ireland, furze (gorse, *Ulex europaeus*) was planted as a crop from at least the seventeenth century not just in hedges but as acreages (Lucas 1960), and in Iceland, lyme grass (*Elymus arenarius*) was harvested for its grain until recently (Gudmundsson 1991). In contrast, there are very few, if any, natural ecosystems of this scale of diversity over such wide areas. The lakelands of Canada and the north European plain might qualify, especially at the borders between major ecosystems. Otherwise, ecosystems of such diversity are usually of restricted size and occur at the junctions of the more uniform ecosystems, where they are known as *ecotones*. We have alluded to several: lake edges, savanna/woodland, woodland/alpine grassland and coasts. They are often youthful, with biological communities unfamiliar by comparison with those of less changing ecosystems, and heterogeneous, with sharp gradients of soils and biota. They are also areas of rapid temporal change, with r- rather than K-strategists being correspondingly predominant in the biota. Often, human communities exploit or settle in ecotones preferentially, partly because of the variety of resources and partly because of easy access to the two adjacent ecosystems (see Fig. 13.2).

A Classification of Ecosystems

Ecosystems can be classified in terms of their structural (=architectural) and bioenergetic complexity (=trophic levels and competition), their productivity, and whether or not they are changing. Age is not considered as a separate parameter, though it is embedded in structural and bioenergetic complexity.

1. Complex, productive, unchanging

Tropical and temperate woodlands, marine coral and algal reefs, estuaries and fens are the best examples. They are structurally and bioenergetically complex, and are highly productive and unchanging on a short time-scale, although fens are less so than the others. For humans, the edible biomass is evenly spread among a high diversity of plant and animal species, although in woodlands and reefs a considerable amount is locked up in structural elements (i.e. wood, coral) and available only for non-edible purposes.

2. Complex, poorly to variably productive, changing

These are *disequilibrium ecosystems*. They are changing, they have a high diversity of plants and animals and thus a degree of bioenergetic complexity, and low to variable productivity. The main biomass for humans is dispersed, but with local concentrations. Examples include grasslands undergoing vegetation change, perhaps through changes in grazing pressure, and frequently managed woodland, such as English coppice woods. Disequilibrium ecosystems were probably more common at times of rapid climate change, as at the beginning of the Holocene.

3. Simple, poorly to variably productive, unchanging

Grasslands are the best example. High-latitude, high-altitude, and coastal grasslands are of low productivity; lowland inland grasslands are of moderate productivity. All see concentrations of animals and birds at various times of the year. Saltmarsh in particular, because of the ameliorating influence of the sea, tends to be productive all year.

4. Simple, poorly productive, changing

Sand-dunes, unstable ground with frost-heaving, and steep hillslopes are the best examples. There is a low diversity of plants and animals, and little of general availability for humans.

5. Simple, productive, unchanging

These are ecosystems which are subjected to short-term change, but from which the plants and animals benefit and by which productivity is increased, and which are unchanging in the long term. This includes ecosystems maintained by pulse stability, such as agricultural groupings of crop plants and domesticated animals, as well as upland and floodplain grazings maintained by fire and alluviation respectively.

Changing Fundamentals

Ecosystems change through time, partly as a function of 'real' change in the geographical environment (chapter 2; Fig. 4.1) and partly because people give different meanings (the perceived environment) to an apparently uniform ecosystem. Giving something a name

Box 4.1 Ecosystem changes in the Holocene

We can illustrate the different types of ecosystem by examining changes in a region of temperate climate during the Holocene, the period from the end of the last Ice Age to the present. The early stages see ecosystems of considerable spatio-temporal variability and ecotones common. There is much bare and unstable ground being colonised by a variety of vegetation and animal communities in a constantly changing pattern; species diversity is rising; r-strategies of reproduction predominate. Productivity is low to moderate. Social ungulates occur in dense, large and visible herds and are mobile, occupying different environments throughout the year perhaps over a range of several hundreds of square kilometres. Soils are thin and immature (rankers and entisols) but of increasing biological activity; they reflect the underlying substrates and are azonal. There is much surface water in the form of lakes and shifting braided or anastomosing streams and rivers; wetland habitats like fen are widespread and becoming increasingly nutrient-rich. Peat deposits are virtually absent, especially blanket peat in uplands. The prevailing picture is one of ecosystem variability at the landscape scale, reflecting the substrate and the topography, and constant spatial change. These are ecosystems of our category 2 (p. 54).

The middle stages, before serious interference by farming communities, see the establishment of zonal (climatic climax) ecosystems – soils, vegetation and animal communities. These are ecosystems of our category 1. Social ungulates are reduced in abundance and the animals of the woodlands occur in small groups and are often secretive; K-strategists predominate. The influence of geology and topography is reduced. There is much greater uniformity of soils and vegetation than formerly, with a reduction in landscape/community diversity. Only above the treeline, along the coastal strip, in saltmarshes (our category 3) and changing ecosystems like river edges and shrinking lakes is there open vegetation, belonging to our categories 4 and 5. Structural and species diversity, on the other hand, is high. Lakes and ponds are becoming infilled and overgrown with fen peat, and rivers more narrowly channelled, especially in the lowlands where extensive reedswamps develop; these wetland fen environments are often highly productive. The prevailing picture is of landscape and community uniformity, high structural diversity and ecosystem complexity, and high productivity. Ecosystems are generally unchanging and there is a reduction in the area of ecotones.

In the later stages, various processes such as soil and vegetational senescence, climatic change and especially the activities of farming communities, lead to a reversal of these trends and a tendency towards the conditions in the early stages. Clearance of woodland arrests the cycling of nutrients and water, and thus leads to a reduction of soil structure, with the result that the effects of geology and topography begin to re-assert themselves. Soil diversity increases (eg Fig. 3.5), with the development of leached, gleyed and eroded profiles; the products of erosion form colluvial and alluvial fills, and rivers become even more canalised as a result. The growth of peat becomes widespread, with blanket peat on uplands and raised bog in valleys and lake basins, further reducing areas of open water. Species diversity falls, although this loss is offset by the introduction of new species by human activity. In contrast, landscape diversity increases enormously. This was the period when some characteristic landscape blocks such as heathland, chalk downland, limestone pavement and moorland emerged from the anonymity of the middle stages (see p. 35). There is considerable ecosystem change and ecotones increase in area. The situation is reverting to that of our category 2 ecosystems.

The latest stages, in urbanisation and large-scale farming, see a reduction in landscape and species diversity, and the spread of simpler ecosystems. Drainage and infilling reduce wetlands, and, generally, land management and planning sharpen gradations between ecosystems and thus destroy ecotones. These are ecosystems of our category 5.

based on the 'geographical' or outward physical appearance can mask diversity of meaning and use.

Marginal land serves as an example. Conventionally land of low soil fertility in relation to other land in the region, it can be used because of economic pressures like war, or expediency as with its nearness to metal-mining areas, or where people are displaced from better lands. It can be used as part of the land of a community for specialised purposes like hay meadow. It can be used by specific kinds of community as a complete system of land use by choice, as with Cistercian sheep farmers in the British Isles. In some cases, immigrant communities may have their own attitudes to the land and not perceive it as marginal at all (pp. 15–16). There may also have been hidden uses: planting and managing boundaries with trees and shrubs for timber, brushwood and fruits, planting ditches with reeds for thatching and stocking them with fish, or using enclosure as a symbol of control. The fields may have been less important than their edges. Alternatively, the land may not originally have been marginal, being used as a normal part of agriculture, but became so through declining fertility, as related to soil or climatic changes or depletion of fuel. Or it became marginal in relation to population decline, technological changes allowing the cultivation of heavier soils in more congenial areas, or changing human values such as lower toleration of access difficulties to isolated valleys. The geographical environment was unchanged, but the perceived environment became marginal. For archaeologists, marginal land by any of the above definitions has common properties of good spatial preservation because of the lack of later intensive land-use, but these mask a diversity of function and perception in life.

At certain times in the past, ecosystems no doubt existed which would look very strange to our modern perceptions. For example, the Late Pleistocene necessitated wholesale adaptation and adjustment of niches by many species, including humans, and allowed colonisation opportunities on a scale not seen since. Earlier, at the Plio–Pleistocene boundary, archaeology has to contend with hominids which probably occupied niches, and had patterns of sociality and behaviour, unlike those of any extant species. In fact, given the predominance today of urban settlement, global atmospheric changes, and high rates of extinction, the present may be a highly eccentric viewpoint from which to study past ecosystems.

Part Two

FROM DEATH TO BURIAL

In Part I, living systems were discussed. All materials, from an individual organism or grain of sand, through soils and sediments, right up to ecosystems, were shown to embody information pertaining to various scales of time and space, scales which are often drawn from, and thus related to, areas and times well beyond the physical limits of the materials themselves.

Now, in Part II, we discuss the transposition of ecosystems into the archaeological record. For organisms, this is *taphonomy* (Efremov 1940), but the transposition of abiotic components of the ecosystem must be considered in the same way (Fig. II.1). At every stage previous relationships are altered – at death there is loss of mobility, at deposition there is generally loss of organic material, and in the post-deposition stage there will usually be physical and chemical alteration (*diagenesis*). Understanding these processes helps us to rewind the sequence of formation of the archaeological record. However, terms like pre- or post-deposition must be used with reference to a specific item in a specific use

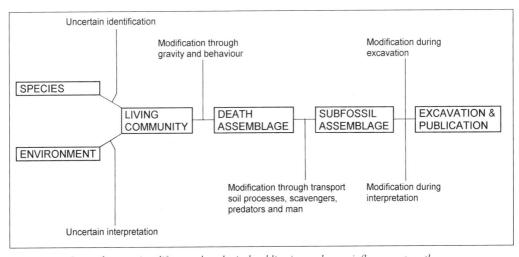

Figure II.1 Stages from ancient life to archaeological publication and some influences upon them.

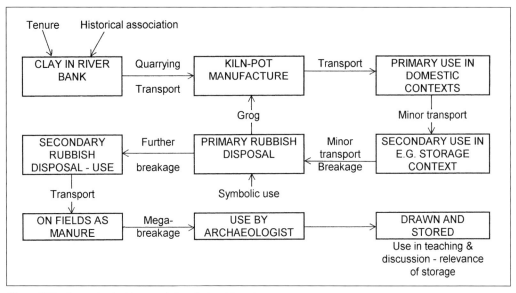

Figure II.2 Stages in the history of the relationships of a pot.

context. Post-deposition in relation to one activity may be a deposition event in relation to another. Transportation of fragmented animal bones from a carnivore kill to a cave is deposition in relation to the kill but the beginnings of a new sequence of events in relation to carnivore den behaviour.

We can extend this view to both ends of the progression from life to death to archaeological study (Fig. II.1). At the life end, we saw this in the embeddedness of time in ecosystems: life incorporates histories (pp. 46–7). In one sense, every living organism has deposited within it aspects of the history of its population, species, and environment, to which are added 'post-deposition' aspects as that organism responds to its environment. At the other end, taphonomy must take in transformations that are introduced by us – our culture, research strategies, excavation and retrieval methods, what we publish and how we interpret things (Part IV). Thus even the terms 'taphonomy' and 'archaeological record' artificially subdivide what is in reality a continuum.

We can illustrate these ideas with a look at the production and use of pottery from a valley clay (Fig. II.2). Clay, a river-bank sediment – itself perhaps a product of human land-use through soil erosion and alluviation – is quarried. First, there is the selection of the locality for quarrying, and this is related not only to functional features like proximity to a settlement or industrial outlet and the quality of the clay, but also to ownership of the locality, its past associations with other groups for quarrying, and the possibilities of a particular clay being used for the contacts it gives between communities rather than its potting quality. Can we ever be sure that we have considered widely enough to encompass all the possible reasons for quarrying a particular locality? Then there are the processes of pot manufacture, followed by transportation to various sites, followed by different categories of use and different categories of breakage and/or disposal. But can we separate use and disposal? Disposal on a midden (see pp. 151–2) may be storage rather than disposal *per se*: pot sherds may have been used for tools like scrapers or burnishers, or ground into fragments to be used as grog in the manufacture of new pots. Midden material including pottery may be used as manure on fields which is then ploughed and eroded downslope to be deposited as river alluvium. The cycle repeats itself. Archaeologists use pottery not just for interpreting and dating a site but in teaching and in discussions on how (or whether) to store and conserve large quantities of pots and so perpetuate the life history.

In effect, the taphonomic progression (Fig. II.1) applies as a series of successive, cumulative increments, each stage overlapping the other, lives becoming deaths, data-gathering becoming interpretation, as we proceed.

FIVE

A CRANNOG AND ITS ENVIRONMENT

We begin Part II with a wilful digression. The spatial transformations which occur at death and burial can most readily be understood by reference to a particular type of archaeological site. To exemplify this, we have chosen the crannog, a type of site common throughout northern and western Britain and Ireland, because the relationships between the site and various scales and qualities of environment are particularly clear (Fig. 5.1). This brief survey of a particular settlement type raises a whole series of questions regarding death and deposition which we go on to address in more detail. The purpose of studying the crannog, then, is as an introductory example to give a point of reference for a number of the points which are made in subsequent chapters.

Crannogs are lake island dwellings (Morrison 1985) and are clearly set in relation to different scales of environment, namely the site of the settlement itself and its island, the aquatic locality of the lake and its girdle of vegetational zones (Fig. 4.7), the dry-ground locality of the hillslopes around the lake, and the wider region. A wide range of biological materials is usually preserved owing to the anaerobic conditions, and usually with a minimum of post-depositional alteration.

DIFFERENT CATEGORIES OF MATERIALS

One approach to the site would be to examine the different sorts of materials preserved in and around the crannog (Table 5.1). Some of these are on the site as a part of the site environment, as with lake-edge plant remains, some are transported to the site by natural processes, as with wind-borne pollen and insects, and some are brought to the site by humans. Practically all the material on the site that has been brought in by humans can yield information at two general scales: the site where it was deposited and the off-site areas from which it came.

DIFFERENT SCALES OF SPACE

The approach we adopt here is to study the different spatial scales – site, aquatic locality, land locality and region (Fig. 5.1). This is of greater relevance to understanding the environment and the way in which data are interpreted (Table 5.2).

A CRANNOG AND ITS ENVIRONMENT

Table 5.1. *Categories of data from a crannog, and the environments and scales of space to which they refer*

Data	Environment	Scale
Sediments	Crannog and shore	Site
Macroscopic plant remains	Crannog	Site
	Lake	Site, locality
	Slopes around lake	
	Woodlands (timber)	Locality
	Crop-growing areas (cereals)	Locality
Large animal bones	Off-site cultural environment	
	Domesticated	Locality
	Wild	Locality and region
Insects	Crannog	Site
	Wetlands around lake	Locality
	Broader lake catchment	Locality and region
	Climate	Region
Ostracods and molluscs	Lake	Locality
Pollen	Lake and wetlands around lake	Locality
	Slopes around lake	Locality
	Broader lake catchment	Locality and region
	Climate	Region

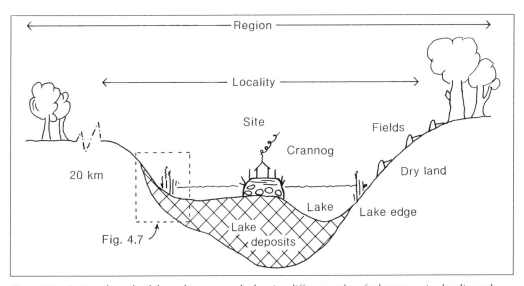

Figure 5.1 *Section through a lake and its surrounds showing different scales of relevance – site, locality and region – in relation to a crannog.*

Table 5.2. *Different kinds of environmental materials of use at specific scales of reference, all from the site*

Site environment
Sediments of the crannog and shore
Vegetative parts, flower and seeds of plants
Some taxa of pollen
Some taxa of insect, especially those associated with specific human activities
Molluscs and ostracods, for the lake and shore at the site
Structures and other archaeological remains

Local environment (This includes the lake, shore and slopes of the catchment)
Seeds, mostly aquatic but some land
Some taxa of pollen
Vertebrates, especially aquatic ones
Bird of prey pellets
Some taxa of insect
Remains of animals and plants used by humans for food
Timber, other building materials, and fuel

Regional environment
Air-borne arboreal pollen
A few insect taxa
Land animals hunted by humans
Materials brought in from the region by humans

The site

A crannog was usually made of stones built up above the water surface to form an artificial island in a shallow area of the lake. A timber palisade was set around the island and there was sometimes a landing stage and causeway to the shore, also of timber. The island floor was levelled with earth, mud and brushwood, and buildings of timber with daub walls and rush or reed roofs were constructed on it. The remains of these structures tell us about construction methods and the spatial distribution of activities on the site. Dendrochronology of structural timber and, exceptionally, log boats associated with the crannogs can date periods of construction, use and refurbishment and may indicate the time of year in which these took place (Baillie 1993).

Sediments at the edge of the island, both those that formed before its construction and those that buried it after it fell into disuse, yield a variety of data at the site scale. Cores or excavations reveal a succession of lake clays and peat reflecting past environments of lake and land (Walker *et al.* 1993). Lake deposits contain molluscs, ostracods and fish, which

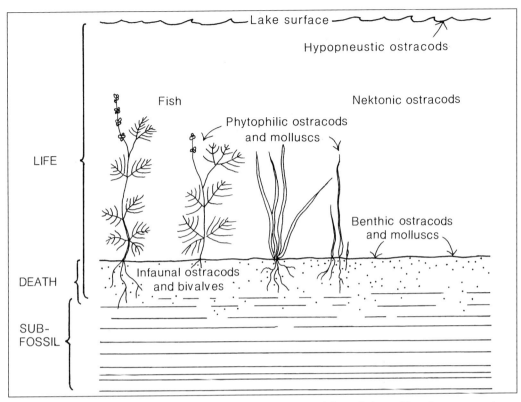

Figure 5.2 Section through a part of a lake and its deposits showing different habitats. Note the occurrence in the upper part of the deposits of both living organisms (the infauna and the plant roots) and dead ones (pp. 57–8). In life, ostracods can be divided into those that live at the water surface (hyponeustic), in the water column (nektonic), on water plants (phytophilic) and on the lake floor (benthic). At death, these distinctions are lost.

indicate the lake environment but at slightly different scales and for different parts of it. Molluscs mostly indicate the nature of the bottom in the littoral zone; ostracods indicate both the bottom and the lake waters since they are swimming animals as well as bottom-dwellers; while fish indicate a wider area than these other two groups since they have a wider range (Fig. 5.2; chapter 11). Diatoms may also give us a detailed indication of water conditions. Macroscopic plant remains, mainly twigs, bud scales, leaves and seeds, reveal the nature of the site and lake vegetation and the stages in the hydrosere (see Fig. 4.7) as the lake dried. Pollen can enhance this picture: species of the lake-edge flora may not all have contributed to the plant macrofossil assemblages. An important category of biological remains at the site scale are insects. Some give information about activities and habitats like leatherworking, grain storage and the state of structural timbers. Others complement the botanical data for the surrounding environment.

So a number of categories of data – mostly of a macroscopic kind like timber, smaller plant parts, molluscs, ostracods, fish and insects, but also some taxa of pollen and diatoms – yield information about the environment of the site and the activities that went on there.

The locality

The locality can be subdivided into three categories: the lake waters, the shore (the lake/land ecotone), and the dry ground around the lake. The main on-site indicators which reflect these areas are animals and parts of plants that actively move around or are dispersed several tens or hundreds of metres rather than those which stay put or move only short distances. The main categories are small vertebrates (fish and various aquatic mammals like otters and beavers), seeds that are wind-dispersed, much pollen, and flying insects. Sorting out which species are relevant to which area of the locality is based on their ecology. Ostracods, molluscs, diatoms and the macroscopic remains of aquatic plants can reflect aspects of the lake locality as a whole, like lake levels, water quality and the nature of the shoreline, but because of their low mobility it is difficult to say how wide an area they reflect beyond this.

An important local category are materials that are not themselves mobile but are brought to the site by other organisms. Beavers move timber around, birds of prey transport their insect and vertebrate victims several kilometres from the hunting site to the roost where the hard parts are regurgitated as pellets, and humans bring all their food – grain, wild plants and meat – from areas where it is produced or collected around the lake and further afield. There are also the building materials of the crannog, most of which are probably of local origin coming from the lake shore, like stone, mud, reeds and rushes, or the immediate terrestrial hinterland, as with stone and timber. All these give us information about the local environments, the dry ground around the lake and beyond, about quarrying and harvesting methods, and woodland management.

The region

At a regional scale, pollen of tree species (arboreal pollen), which tends to disperse further than that of shrubs and herbs, is the main indicator. This tells us about the regional woodland composition. The ratio of arboreal to non-arboreal pollen (AP:NAP) can be an indicator of the degree to which the land was forested or open, but since much of the non-arboreal pollen is of local origin, coming from non-climax aquatic communities around the lake edge (Box 6.1), this ratio must be treated with caution as a commentary on the regional environment. Usually, the non-arboreal pollen is separated into ecological categories such as wetland, dryland and arable, in order to fine-tune this category. Another group in the regional category are animals and plants brought to the site by humans and other animals. Human hunters and gatherers tend to range further than farmers, so the species brought in to the crannog from these activities are an important guide to the environment beyond the farmed area. Items brought in by trade or exchange may have originated many days' travel away from the crannog. A few winged insects might also be of regional significance.

The wider environment

Some species of plant and animal are of climatic significance, telling us about changes of temperature and rainfall. However, insects that are favoured by human habitats must be interpreted carefully in this respect since the exceptional conditions of the site may allow their presence in areas from which they would otherwise be absent, especially if at the limits of their overall range. Volcanic ash layers are also sometimes present in peat and lake sediments and these can be useful as chronological markers, especially where they have a clear geochemical fingerprint.

RELEVANCE OF THE SPATIAL INTEGRITY OF THE DATA

We have shown above how data pertaining to different scales of environment, from the site to the region, can be obtained from a single, theoretical sample or, at most, a site. This is, however, an early stage in research strategy, for although the data are good in terms of their sample or site specificity, they are lacking in detail at the wider scale. High-resolution information is obtained about activities at the site and intra-site scale – precisely where the crannog was located, stakes and wattling palisades constructed, houses built and perhaps the uses to which rooms, working areas and storage pits were put. There is little discrepancy between the original 'life range' of the materials and the place where they are preserved. On the other hand, as indicators of the same sort of detail within the locality and region, the data are poor. With some resources, like mineral ores or rocks which have narrow geological signatures, sources may be pinned down precisely; with others, as with the feeding grounds of animals, they are often vague. The general nature of the lake shore, whether there are extensive reed beds or stony beaches, and the different types of agricultural land, whether arable, pasture or reverting to scrub, are indicated, but not the detail of where those shores, arable land and woodlands were. For this we need field survey, and this would be one of the next stages in setting the crannog into its wider environment (chapter 9). There is thus a contrast in the on-site data between that relevant to the site, which gives high spatial resolution, and that relevant to the locality and region, which is of low spatial resolution but which is of a more general relevance.

THE SPECIAL NATURE OF THE SITE

In considering the data, the location and nature of a site must be taken into account. All environments, as well as reflecting general conditions of the region, as with climate and vegetation, and the specific environment of the locality and site, as with the lake and crannog, are particular in more subtle ways. In large lakes, fauna and vegetation can be segmented into geographical areas corresponding to topographical units like bays, each with its unique associations of species and variants (pp. 18–21), which relate to their isolation from adjacent communities as much as to other aspects of the physical environment. Foci of human activity are not randomly placed, but relate to the physical environment, culture and society, interactions between these, and deeper aspects of the human psyche. Crannogs were often sited in sheltered, shallow areas of a lake, so, along with the local biogeographical variations just mentioned, lake assemblages from the site relate to both shallowness and isolation. Sites can also reflect the wider environment in special ways. Thus crannogs were sometimes dwellings of high status, even royal, people, and it can therefore be expected that the materials of which they were built and the food of their inhabitants were different from those of other people. Building skills and materials may both have been exceptional, and the subsistence catchment of the site may have far exceeded that of contemporary peasant farmsteads (Campbell and Lane 1989; Lynn 1983). It may also be that royal crannogs were placed in exceptional locations, for example, especially sheltered and isolated bays, with the result that the exceptional nature of the lake environment at a crannog site was further enhanced. So the special features of sites, both cultural and environmental, have to be considered before drawing conclusions about the general environment, although, of course, this variability is a part of that wider picture. In the next two chapters we will see that only some sites are preserved and that this is not a random sample of the original distributions but relates partly to properties of the site, thus creating further heterogeneities in the data.

The Site Catchment

All the archaeological materials and documentary sources from the site or pertaining to it allow the definition of the area from which the site obtained its supplies, whether these be day-to-day victuals like milk, grain and fish, the more occasional materials like timber and game, or commodities coming from much further afield. This entire area is the site catchment (SC), the area of total exploitation as based on the archaeological evidence of known resource use (Bailey and Davidson 1983). Immediately adjacent to the site, exploitation can be intensive; further away it is less so, as with the grazing of domestic flocks and herds, hunting, or the quarrying of inorganic raw materials. Catchments sometimes incorporate detached localities where there are critical resources like hay meadow or woodland, and there can be overnight or longer stops in subsidiary camps or dwellings (e.g., see pp. 190–1).

In this chapter we have shown how the study of a site can give information of a variety of types and at a variety of spatial scales. The study does not stop here, however, and in successive later chapters we show how other methods can amplify these data and, especially, how they can pin down the localities where various activities took place away from the crannog.

Six
DEATH ASSEMBLAGE FORMATION

DEATH OF ORGANISMS

At death, an immediate change occurs: organisms cease to move under their own power. Of the many loci inhabited during life, only one is represented at death, and that no larger than the corpse (Fig. 6.1). Furthermore, the death location is often unrepresentative of any of the life situations. If we find a heap of dead snails in the garden, it is not possible to say from that death assemblage alone whether those snails limited their activities to the immediate area or went ramping throughout the garden and into the neighbour's cold-frame as well. There is a significant difference between human and animal death

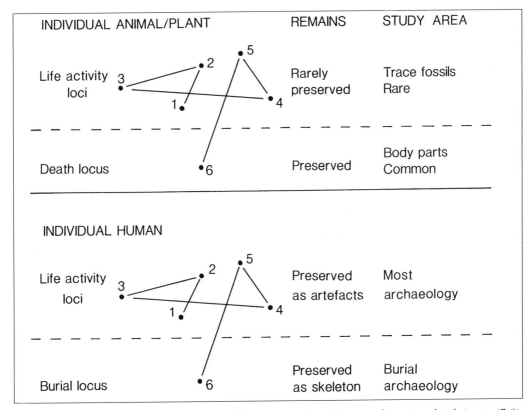

Figure 6.1 The contrast between animals/plants and humans in regard to their preserved remains and study (see pp. 67–8).

assemblages (Fig. 6.1). Humans generally dispose of their dead in an ordered fashion in particular places set aside for this purpose, so most of their activities are represented by their material culture rather than their bodily remains. For animals and plants, the reverse is usually the case, with the bodily remains rather than their constructions or other traces being the main data of study, although there are exceptions such as footprints and faeces, and the wider effects of soil animals.

Attritional death

Several processes can lead to the formation of a death assemblage. The most obvious is natural death, from juvenile mortality, disease and senescence, which leads to the attrition of a population. This can reflect the mortality profile of the population and thus the population structure in life; K- and r-strategists, for example, should show different profiles (see Box 3.1). However, this connection can be compromised by the relationship between sampling location and the behaviour of the organism in life. The degree to which the death assemblage differs from the life community relates to the mobility of the organism in the first instance (pp. 21–3). For example, some large vertebrates are dispersed for the whole year, and it would be virtually impossible to establish the mortality profile of their populations from skeletal remains alone. A death assemblage of rock-encrusting organisms on a seashore or of parasites in a mammalian gut varies little from the life community. For multi-locational organisms, the attritional death assemblage is spatially dispersed and the overall mortality profile not represented at any one sampling spot. A breeding shearwater colony (see Box 3.2) has a death assemblage of shearwater corpses, but this represents the mortality of the population for only a part of the year, and from a particular source – gull predation. During the rest of the year when the population is more dispersed, deaths are under-represented in the breeding area so certain age groups are absent from this particular death assemblage. Some organisms, notably plants, have components which are unilocational – the breeding plant – and others which show varying degrees of dispersal – leaves, seeds and pollen. The different components form death assemblages which have different characteristics. An oak tree throughout its life generates assemblages of pollen, which are deposited close to and far away from the tree, and assemblages of seeds, which are less widely dispersed. Each represents a short period of the life of the organism. When the tree itself dies it constitutes a death assemblage which represents the unilocational part of the organism and its lifespan.

In terms of information, there is an increasing area of reference but a decreasing degree of detail as we go from sessile organisms to mobile ones (pp. 21–3). The same applies in the sequence from precipitates, through soils to clastic sediments (Tables 5.1 and 5.2, and Fig. 7.6).

Death-traps or locationally biased death

There are places within the life range of an organism where it is more likely to die than others. If we were looking for dead flies in a house, the windowsill on the inside of a sunny window would be a good place to start as the combination of environmental factors at such a point is likely to attract and kill large numbers. The tar-seeps at Rancholabrea, California, have trapped a substantial death assemblage of mammals because they are places where death is more likely to occur, thus concentrating the attritional death assemblage in one part of the range. Pits serve the same function, and assemblages of small vertebrate bones and teeth in storage pits and wells on archaeological sites are common. The chances of death and death assemblage formation are not equal across the range of an organism. Moreover, these are not random samples of the wider community, since they have their own special environments which attract and select only certain groups.

The effects of predators and scavengers

Some death assemblages are produced and modified by predators and scavengers. Selective predation may have taken place to optimise energy return in relation to finding and handling time, or for palatability or visibility. Predators of vertebrates may partially dismember corpses at the kill site and remove some parts to a lair, creating highly modified death assemblages. These may be more informative about the activity of the predator than about the range and community structure of the prey. Thus the number of specific skeletal parts in bird-of-prey pellets can be used to identify the bird species and its hunting range and environments (Andrews 1990; Yalden and Morris 1990). Scavengers disrupt death assemblages produced by attrition or predators and remove parts of them. A hyena may rummage around the gnu death assemblage formed by a lion, scrunching up and totally destroying some of the bones and then removing others and redepositing them in its lair. This can be selective for certain bones and parts of bones as related to their size, resistance to chewing, their adhering meat, and their marrow and grease (Box 6.2). Thus scavengers and predators relocate and concentrate selected parts of the prey death assemblage and leave a depleted, but equally non-random, assemblage at the kill site.

Catastrophic death

Vertebrates may undergo catastrophic mortality, in which a specific event such as a fatal epidemic disease brings about the death of large numbers of individuals. Such an event is unlikely to be arbitrary in the age and sex groupings which it generates, as some sections of the population may be more susceptible to the lethal disease, or only a certain part of the population may be located within the 'killing zone' of a lethal event. Catastrophic mortality death assemblages are markedly different from attritional ones.

Hominid- and human-originated death

Human food debris, likewise, is never representative of the full biological community within the operational environment of the human population. There is selection of species, age, sex, health and condition groups, and humans transport food beyond its life range. Human predation may tend toward the attritional or catastrophic mortality types, killing small numbers of a population over a long period, or causing slaughter of large numbers

Box 6.1 Modelling pollen assemblages

Pollen grains are transported in various ways. Some species of plant release huge amounts of pollen to be transported by air currents, whilst others have means of attracting insects and other vectors in order that they will transport pollen from one flower to another, and so produce relatively little pollen. In some plants, like wheat and barley, flowers do not fully open and so only negligible amounts of pollen are released. Much of the modelling of pollen deposition has focused on the movement of pollen in air currents, and on understanding the sources which contribute to polleniferous sediments (Moore *et al.* 1991).

Tauber (1967) considered a small open mire in wooded land, and proposed different components to the pollen rain, each originating at a different distance, and each subject to varying degrees of modification of the original pollen output. The local component may be dominated by sedges (Cyperaceae) and *Sphagnum* species, plants growing on or immediately adjacent to the point of deposition. Within the surrounding woodland, air currents circulating within the trunk-space transport pollen from ground and shrub vegetation, whilst the component derived from the canopy may be dominated more by pollen of trees. Long-distance transport brings pollen from plants growing many kilometres from the point of deposition, especially if those plants are wind-pollinated. The significance and content of the last component depends on wind speed, prevailing direction and precipitation. Research in the Shetland Islands has confirmed the transport of pollen over tens to hundreds of kilometres, given appropriate conditions (Tyldesley 1973). Pollen is also introduced from plants growing within the catchment in streams which pick it up during flooding and transport it onto the mire. This offers a means by which pollen taxa not normally widely dispersed are transported some distance. Streams also derive pollen from eroding soil or peat which may be centuries or millennia older than pollen freshly deposited into the stream at the same time. In

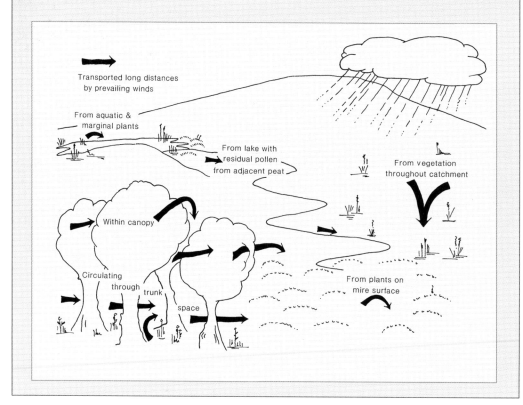

Modelling pollen assemblages

almost any polleniferous deposit, then, there is likely to be local pollen derived from the contemporary vegetation close to the point of deposition, regional pollen from long-distance transport, principally by air and water, and a residual component of reworked pollen of various ages.

Understanding the spatial scale of pollen deposition has become vital as pollen analysis has shifted from reconstructions of regional vegetation to local studies (Jacobsen and Bradshaw 1981). Size and type of pollen-collecting site are important. Large basins have a catchment of several square kilometres, whereas small bogs receive most of their pollen from a few tens or hundreds of metres away; different-sized basins 'sense' the environment at different scales (Barber 1988). The underlying principle of all modern studies is to compare the pollen rain at a particular spot with surrounding local and regional vegetation, and to generalise on the movement and deposition of pollen and the spatial catchment of the point of deposition.

At the largest scale is the European Pollen Monitoring Programme (Hicks *et al.* 1996), set up 'to use pollen traps to monitor pollen deposition across vegetation transitions from closed forest to open situations . . .'.

Janssen (1966) studied vegetational transects in Minnesota, across belts of woodland and scrub to emergent vegetation at the edge of a lake, and on to open water. Some species showed peaks of pollen deposition at the point where that species was most abundant. The maple *Acer spicatum*, for example, was prominent in pollen from surface samples only within the relatively narrow zone where that tree predominated. Pollen of other species was present throughout, notably of the wind-pollinated pine and birch. Pollen from the mid-point of a 600 m-wide lake was dominated by pine, birch, grasses and alder, with little pollen from aquatic plants, showing that inter-species differences in production and deposition can mask spatial detail. Above all, lateral movement of pollen blurred the zonation of lake-margin vegetation.

Research by Hall (1989) on moss polsters raised the important point that small-scale variation in the surface onto which pollen falls influences preservation. In a study of one year's pollen rain on grazing land in the north of Ireland, pollen falling onto moss polsters was significantly better preserved than that which fell onto grass or bare soil. Grass-covered surfaces also trapped pollen more effectively when grass growth was at its most vigorous, in late spring and summer, thus biasing assemblages towards taxa producing pollen during those seasons (Hall 1989, 66–7). In the same part of Ireland, Gennard (1985) also underlined the importance of different levels of pollen productivity by observing that pollen of flax (*Linum usitatissimum*) is scarce even on samples of flax stems and bolls from a contemporary crop.

By combining data from catchments likely to represent a large region with data from much more local catchments, Bradshaw (1991) and Mitchell (1988; 1990) were able to investigate woodland history in south-west Ireland in greater detail than would have been possible from the big regional catchments alone.

Studies of modern pollen catchments have thus been of value in refining the interpretation of ancient pollen data. In particular, they have shown the considerable degree of 'distortion' which occurs between the living population and the death assemblage, and the way in which some species may be almost invisible in the death assemblage. Much of this work has been empirical: despite Tauber's original work, surprisingly little formal, quantified modelling of pollen assemblage formation has been done.

Table 6.1. *Summary of types of animal death assemblage formation*

Attritional	Sedentary, non-dispersed animals – mortality profiles a close reflection of population structure in life
	Mobile, highly dispersed – mortality profiles a poor reflection of population structure in life
Catastrophic	Death assemblage closely related to death environment
Locationally biased (death-traps)	Death assemblage related to animal behaviour
Predator-mediated	Death assemblage modified, dispersed and reduced
Scavenger-altered	Death assemblage differentially reduced and dispersed
Hominid- and human-originated	Death assemblage modified and dispersed according to economy, e.g. scavenger or urban dweller

of individuals simultaneously. The sequence of locations from live animals through bones to ultimate burial (Box 6.2) is complex, even with hunter-gatherer and subsistence farming sites. With villages and towns there can be many more stages between the original husbandry practices and the depositional assemblage in the archaeological record.

These different kinds of death assemblages are summarised in Table 6.1.

Modelling Death Assemblages

The course taken by different groups of organisms from the biosphere to the lithosphere has been modelled for each of the many stages of this taphonomic progression. For pollen, we have chosen to look at the sources and distances from which pollen is derived. For bone, we look at a later stage, namely the local histories at and after death (Boxes 6.1 and 6.2).

Community Death

For multiple-species death assemblages, the various components may have arrived by different means and represent a variety of types of death. In a death assemblage of mixed taxa of varying mobilities and ranges the life population structure is represented in only a few of these, generally the less mobile. This is an argument for considering death assemblages which mix vertebrate, invertebrate and plant remains as a series of quite distinct assemblages (e.g. Table 5.1).

For complex communities such as woodlands, death results in a reduction of the three-dimensional structure to two dimensions on the woodland floor, just by the processes of gravity and decay. For example, suppose that three insectivorous bird species, A, B and C, live in the same area, even on the same tree, but at different heights (Fig. 6.2); C in the herb layer, B on the trunks of trees, and A in the leaf canopy. In area 'X' (the tree), A is present all year round, B is a summer visitor from area 'Y', and C is a winter visitor from

DEATH ASSEMBLAGE FORMATION

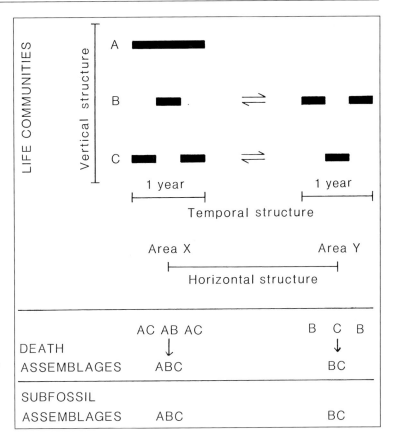

Figure 6.2 The decrease in temporo-spatial structure of an animal community from life through death to burial. See pp. 72–3 for explanation.

area 'Y'. This vertical and annual-timescale horizontal structure is lost at death, and even the seasonal death assemblages, AC, AB, B and C, are rapidly combined into ABC for area 'X' and BC for area 'Y'. Likewise with lakes (Fig. 5.2), and unstable habitats like colluvial surfaces (Fig. 7.6), vertical and annual partitioning is destroyed at death.

On river floodplains, there can be two seasonally separated groups of species, aquatic animals during winter flooding and land animals during summer dryness (Robinson 1988). Survival of each group in the adverse season depends on the ability to lay desiccation-resistant eggs, or burrow and remain dormant, or move away and then immigrate in the favourable season. The close juxtaposition of ecological groups is most apparent in complex environments like swamps where there are micro-habitats in close proximity – sedge tussocks which are virtually terrestrial all the year round, pools which are aquatic all the year round, trees with a variety of epiphytes, and fallen branches (Bishop 1981). At death, all of these become mixed, due to gravity and slight lateral movement caused by wind and currents.

Where the death assemblage is a more or less true representative of life it is referred to as being *autochthonous*, but where there has been some mixing the term *allochthonous* is applied. These terms must be applied with reference to particular assemblages and processes. Here are some situations:

Box 6.2 Modelling bone death assemblages

Bones are dispersed only after death, although the mobility of most vertebrates means that death locations are varied. Vertebrates share interactions with humans during life, so modelling bone deposition is closely involved with modelling human hunting and butchering (Shipman 1981; Lyman 1994; Stahl 1996). In life, a mammal bone is one of about 200 in the body. From its origin in the foetus, a bone grows, has denser and softer parts, fuses with other parts, varies with sex and age, is modified by environment, and takes up chemical signals according to diet. Killing and butchering involves selection, e.g. for fat, pelt, health or age. At a kill, if there is competition from scavengers or others of the same species, or a need to feed young elsewhere, parts of the carcass are removed. Each bone can be acted upon by carnivores that slice the meat off it, scavengers that crunch it to get at the marrow in the shaft cavity and the grease in the epiphyses, animals that swallow it and corrode it with gut acids, humans that modify it for a tool, feet that trample it on and into the ground, decomposers that act on it biochemically, and soil conditions that destroy it, although a geochemical signal may survive. Bones taken to another site may be the best edible parts or have been selected for other uses. Bones of best meat and marrow quality are often those destroyed, while the least nutritious are preserved (Binford 1978; Jones and Metcalfe 1988). Food-sharing (Marshall and Pilgram 1991), use of rubbish for particular purposes (Hodder 1982a), and deposition of bones of particular types in certain contexts (Hill 1995) extend the use of bones beyond subsistence alone.

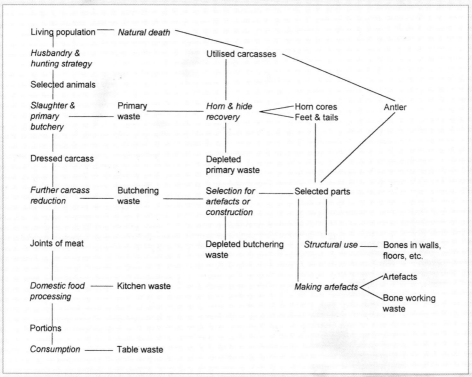

Hypothetical flow diagram to show stages in the transformation of a population of living vertebrates to bone assemblages. Processes are in italics. After O'Connor (1993).

Modelling bone death assemblages

In sorting out these processes, a variety of approaches is used: ecology, experiment, observations in the wild and ethnography. Ecology provides initial clues, for it is likely that some species were occupying a site while others were not. Fissures and relatively inaccessible caves are not usually lived in by ungulates but are by carnivores, so ungulate bones probably indicate carnivore or human activity. However, ungulates and other herbivores enter caves occasionally, as at Mt Elgon in East Africa where animals, including elephants and buffalo, go for salt (Sutcliffe 1985).

Separation of human and animal, and carnivore and scavenger activity has been studied by experiment. Payne and Munson (1985) fed a known number of animals to dogs and examined material not ingested and in faeces. Crushing and corrosion selectively destroyed weaker bones and those of smaller animals: survival was high for teeth and early-maturing long-bone ends, moderate for metapodials and podials, and low for late-maturing long-bone ends, scapulae, pelves and phalanges. Differential survival on archaeological sites may thus be a function of scavenger activity (selective destruction) rather than selection of particular joints or bones by humans for particular uses (selective transport). Experiments with flint or iron knives on bone surfaces of different densities, age and processing (e.g. roasted, boiled or uncooked) have produced marks that can be compared with those from archaeological contexts. Tooth-marks of animal carnivores and bone smashed by human hammerstone activity have been used to assess the primacy of these two agencies. Blumenschine (1988) compared the effects of spotted hyenas on bone that had been smashed with hammerstones and the marrow extracted and unsmashed bone. In the smashed bone group, hyenas removed some bones because of competition, all the epiphyses (articular ends) were eaten for their grease, and 15 per cent of the remaining shafts was tooth-marked. In the unsmashed bone where the marrow had not been previously removed by humans, 82 per cent of long-bone shafts bore tooth-marks. So here is an index, namely the different intensities of tooth-marks and hammerstone breakage, to test the primacy of hominids over carnivores in forming bone assemblages. However, some bones tend to be discarded untouched, like metapodials that are low in grease and those of stressed animals that are low in fat. Habitat of bone discard is important too, hyenas only lightly crunching or totally ignoring bones in some areas, while in other areas there is repeated activity.

Observations on African elephant bone under non-cultural conditions showed 'cut-mark mimics' created by trampling (Haynes 1988). However, Olsen and Shipman (1988) claim that the surfaces of experimentally trampled bones and those which had their soft tissues removed with a flint tool showed significant differences. Studies of Okiek hunter-gatherers of Kenya showed meat-sharing to affect body-part representation in the bone assemblages and to create similar profiles to those of differential transport of carcass parts (Marshall 1994). Sharing occurs between households and is chiefly patterned by social relations, hunting success and animal size, with households of successful hunters and well-liked individuals accumulating more bones and bones of higher nutritional utility. The effects of carnivore activity have also been studied by looking at droppings collected in the wild, although a problem with this approach is that we do not know the original condition of the prey or what was actually eaten so we cannot quantify the effects. Stallibrass (1990) contrasted fox droppings, in which bone was fragmented and corroded, with owl pellets, in which bones are often complete.

Similar work has been undertaken on fish bones. In feeding experiments with dogs, pigs, rats and himself, Jones (1990) showed that some bones survived better than others, although overall destruction was greater than in comparable experiments with mammal bones. In an ethnographic study Stewart (1991) separated natural death assemblages and fishing-camp assemblages at Lake Turkana, Kenya, with fishing-camps having lower taxonomic abundance, medium size of fish, higher bone frequency, cutmarks and cranial fragmentation, and more axial bones. Zoologists studying otter droppings to monitor the species and sizes of fish eaten have shown that this form of fish death assemblage has distinctive characteristics (Jacobsen and Hansen 1996; Carss and Parkinson 1996).

1. Autochthonous with reference to life and death: the biota live on and form the sediments in which they are preserved, as with coral and peat.

2. Gravitationally allochthonous with reference to life, autochthonous with reference to death: mixing is due mainly to gravity, as in woods (Fig. 6.2) and lakes (Fig. 5.2).

3. Behaviourally allochthonous with reference to life, autochthonous with reference to death, as on river floodplains and in swamps.

4. Depositionally allochthonous with reference to life and death: this is where there is lateral movement, slight in swamps and fens, considerable in river channels. (Strictly, this belongs with deposition (chapter 7), but since it is a part of death – sometimes at the moment of death (as with rivers or bird-of-prey kills) – it is included here.)

Ecosystem Abandonment

In some cases, ecosystems are abandoned by people. This can involve intensifying activities in a smaller territory, as with a shift from extensive agro-pastoralism to intensive arable farming, or deserting entire settlement systems. The idea of abandonment helps us focus on the very last activity of a site or region, especially important since this may have been atypical of the activities which took place over a previous and longer period (Cameron and Tomka 1993). For example, there may have been a long period of settlement in which houses and enclosures were of timber and a very short final phase when, in order to strengthen them against a perceived threat, they were of stone. In areas which are now marginal land, obviously the latest episode of activity before abandonment is the most visible, but this may be just that episode which either led to the marginality of the land or was a response to its marginalisation. Much of archaeology is about these atypical abandonment episodes.

Abandonment (usually implying human abandonment) refers to particularly striking changes; however, life continues in the 'abandoned' areas. Not all of the ecosystem dies; it is not so much a death assemblage that is created by abandonment as a redirected ecosystem, the nature of which crucially depends on what was going on just before abandonment. If lightly grazed land is abandoned, leaving a rich and unbroken grass sward, it may be a long time before woodland regenerates because of the inhibitory effects of the grass sward to seed germination. If there is overgrazing at the time of abandonment, scrub and woodland, being able to colonise patches of bare soil, have a better chance of taking over. If overgrazing is extreme, abandonment may lead to erosion, as with the collapse of field boundaries and terraces (see Fig. 10.8 and pp. 129–31). The nature of abandonment is also important. Settlements which are abandoned suddenly and totally and the buildings and gardens left will have a much richer fauna and flora, albeit a particular one of synanthropic species, than those where dwellings are dismantled and all traces of human occupation removed.

Ecosystem Death by Burial

Often ecosystems are killed by the processes which preserve their remains, and this is usually by burial of a land surface under a sediment, although it may also apply, for example, where rising lake levels preserve a crannog through waterlogging. A key question is: 'What is the relationship between the former ecosystem and the cause of death and burial?' This is relevant to the interpretation of the ecosystem in relation to a wider area: are we dealing with a random sample or is there something about the area that has been preserved which relates to the mode of death and burial? Several categories can be suggested:

1. No relationship between life and death. This is the situation with unique sudden events of a regional or wider scale, like meteor impact, tidal waves (Fig. 1.1) and volcanic eruption, which elicit no response from the ecosystems and the people in them, either before or at the time of death (but see Gould 1980, for adaptations to tidal waves).

2. A general and specific relationship between life and death. Death and burial are caused by a sudden event which is inherent to the life environment, like a sand-storm, flash-flood alluviation or rising water levels. In the case of a crannog (chapter 5), defence in a lake may have made it attractive to human occupation in the first place, as well as vulnerable to flooding by a rise in water level. If the event is expected, specific plans could be in place to minimise damage, as with the construction of ditches and the raising of settlement mounds in areas subject to flooding (Brown 1997, 304–16).

3. A specific relationship between life and death; burial is sudden. This is the situation with human constructions, the location of which is almost always related to the modified environment (Fig. 2.3), and especially to previous land use, as with Neolithic and Bronze Age monuments in southern Britain (Evans 1993).

4. A specific relationship between life and death; burial is gradual. The ecosystem gradually swings towards the conditions that ultimately bury it. This is the situation where increasing intensity of sand-storms leads to clogging of harbours, reduced land rentals and partial desertion before burial; or where valley-bottom grassland suitable for all-year-round pasture is flooded, making it suitable only for summer pasture, and then buried by alluvium. In the Upper Thames Valley, for example, there is a dry-ground brownearth soil (Limbrey and Robinson 1988) in the earlier Holocene, and then a sequence in later prehistory of waterlogging, flooding and the use of the floodplain for summer pasture, along with archaeological evidence for enclosures and buildings, all before alluviation (Robinson and Lambrick 1984).

5. A continuing relationship between life and death. Death and burial are an integral part of the contemporary ecosystem, as with the formation of coral and algal reefs and overbank alluviation. Deposition is so fine that it allows the continuation of life and does not change the ecosystem.

In short, it is nearly always the case that understanding the processes of death and burial of a past ecosystem is indivisible from understanding the 'life' of that ecosystem.

Seven

Deposition and Post-Deposition

Deposition is the incorporation of death assemblages and ecosystems into preservational media. *Post-deposition* is used here to mean their reworking or influencing by processes which were not going on when they were first incorporated. The terms must be used with reference to specific events or contexts (pp. 57–8). The processes involved are a combination of extrinsic and intrinsic factors – the environment and what is being deposited. Again, this is an artificial distinction because the materials being deposited are themselves a part of the environment, even modifying it in some cases, while the environment is itself often largely a product of human activities in the first place.

The Nature of the Archaeology

The nature of the archaeology varies in its potential for visibility and preservation (Fig. 7.1). Stone tools are more durable than wooden ones, monuments and buildings more durable than tents, and not all activities are within the resolution of the archaeological record (Fig. 7.2). At one extreme, a human population may, by the use of grazing animals, gradually bring about extensive and long-lasting vegetational change, which in turn may cause marked changes to soil and sediment systems. If the original vegetation cover was patchy, the effects may only be obvious when a large area is examined, so as to 'even out' the smaller-scale variation, but may nevertheless be clearly detectable in the pollen rain, rates of soil erosion, and a replacement of plant-feeding

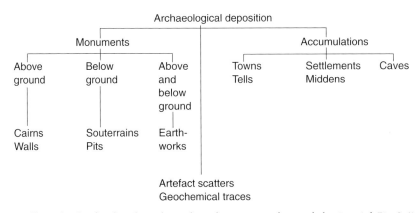

Figure 7.1 Different kinds of archaeological signal in relation to past human behaviour (cf. Fig. 9.4).

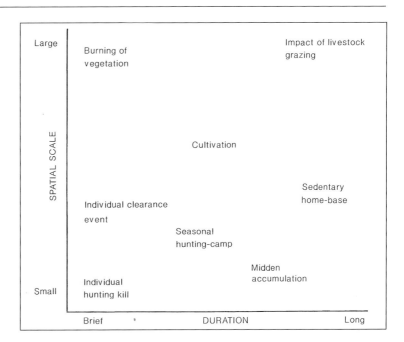

Figure 7.2 Different categories of human activity in terms of time and space.

arthropods by those which infest herbivore dung. At the other extreme, the clearance of an area of scrub by burning will have a substantial immediate effect, evident within hours or days, but possibly almost invisible after years or decades unless the burning is repeated. Similarly, if a hunting people clear a small area for a camp, the effects may only be apparent over a hectare or so, whereas the effects of their predation pressure on surrounding animal populations may be apparent over many square kilometres.

The nature of abandonment is also relevant (p. 76). In the case of the crannog (chapter 5), a site in a shallow and sheltered area of a lake could be abandoned gradually, with the buildings being dismantled and, along with the small artefacts, removed. A site further from the shore and in deeper water might be abandoned more rapidly and much of the materials left at it. If these two locations reflect differences in the type of crannog in life – one royal, the other not, or one a domestic settlement, the other for fowling and fishing – then this might be masked by different patterns of abandonment. The nature of the contemporary regional abandonment or continuing land-use is also relevant to site preservation in that soil erosion through agriculture or the collapse of field boundaries (see Fig. 10.8) can cause material to be washed into the lake and thus enhance preservation of at least those sites in shallow water. Preservation is also a function of the location of sites. In the case of the two different crannog locations, there might be a contrast in preservation, with the more sheltered sites closest to the shore and in shallow water being better preserved by silting than those in deeper water which are more prone to erosion. If these different life contexts saw correspondingly different types of crannog, the surviving data would be slanted towards one type and not the other. When there is a cumulative correspondence between environment, site type in life, type of abandonment, and preservation or destruction, the data become very slanted indeed. And this is before further selection by archaeologists for study (chapter 8)!

Box 7.1 Physico-chemical conditions of preservation

Several widespread preservational environments can be defined with reference to pH and oxygen, together with some situations where they occur and some biological indicators.

Depositional environment	Main soil & sediment types	Some typical situations	Environmental indicators
Acid, pH usually <5.5, oxic	Podsols and other leached soils	Heathlands, upland moors, some river gravels	Soil pollen, charcoal, phytoliths
Basic, pH usually >7.0, oxic	Rendsinas, lake marls, tufa, alluvium, shell-sand	Chalk and limestone areas, valley bottoms, karst, machair	Molluscs, bones, charcoal ostracods
Neutral, pH 5.5–7.0, aerobic	Brownearths and gleys, river gravels	Clay vales and other lowland plains	Charcoal, sometimes bone and shell
Acid or basic, anoxic	Peats and organic deposits, e.g. lake sediments	Urban sites, wetlands, river floodplains	Insects, macroscopic plant remains, pollen, bone, wood, charcoal

Acid oxic environments

Where pH is below 5.5 and the soils and sediments are fully aerated, the environment is acid and oxic (i.e. oxygen is available for direct chemical reaction and for aerobic organisms). The main soil type in temperate climates is a podsol, in which leaching and the formation of strongly horizonated profiles are predominant features (p. 32). The soils develop on nutrient-poor and freely-draining parent materials, often where there is high rainfall, but also in low-rainfall areas where subsoils are poor and where there has been no attempt by humans to upgrade the soil. Heathlands, moorlands and some river gravels, especially on older terraces where there has been more exposure to leaching, are typical areas of podsol formation, and this type of environment is widespread in the less elevated uplands (around 400 m above sea-level) in the British Isles. Organic materials are not generally preserved, although in some cases deposits of organic matter such as conifer leaves accumulate. The acidity destroys bone and shell. However, the actinomycetes which particularly attack pollen are inactive in low pH conditions, so pollen survival may be good. Plant charcoal, the silica bodies of grasses known as plant opals or phytoliths (p. 135), and cremated bone may also survive well. In some cases deposits of the silica tests of diatoms (p. 136) are present in great concentrations especially in lake deposits (diatomite).

Basic oxic environments

The pH is above neutral, with soils and sediments distinctly calcareous and fully oxic. The main soil types include rendsinas on limestone, chalk and coastal shell sand, and calcareous mollisols in areas of steppe grassland. Associated deposits are valley colluvium – ploughwash and cold-climate solifluxion deposits – and, in low-lying areas near springs, tufa. Alluvium, including fine silts and loams and coarser gravels with organic lenses, is often highly calcareous, especially in the upper reaches of

> ## Physico-chemical conditions of preservation
>
> river valleys and lower terraces where there has not been intensive weathering. Karst is typical of harder limestones with features such as caves, fissures and rock pavements. Decay of organic matter is rapid and more or less complete, except where incorporated as humus, while mollusc shells and ostracod valves are well preserved. Bone, including its protein component, is preserved although surfaces are often badly corroded. Caves and fissures are frequently rich in bone. Calcareous precipitates – travertine (Fig. 3.11), stalagmite, tufa – often preserve casts of plants and animals including the algae which formed them.
>
> ### Neutral oxic environments
>
> The pH of neutral oxic environments is between 5.5 and 7.5 and soils and sediments are more or less oxic. Soils are usually of brownearth type with (B), Bt (argillic brownearth), or weakly developed iron accumulation horizons; gleying occurs in partially waterlogged soils. This state of affairs is typical of wide areas of temperate lowlands and is the context of many archaeological sites. Clay vales, some sandstones, the more impure limestones and extensive areas of river gravels support this kind of depositional environment. Biological materials are poorly preserved. Only charcoal is at all common. Occasionally where there is reduced water movement and relatively high pH, bone and shell are preserved. Organic materials are seldom preserved.
>
> ### Anoxic environments
>
> These are environments that have been continuously anoxic, lacking free oxygen, since deposition. Some decomposition may take place for a short while and continuously at very low levels because some anaerobic organisms may be active. These conditions are widespread and varied, encompassing the lower strata of some urban sites, the bottoms of pits and ditches especially in floodplains, blanket peat, wells and shafts, lowland wetlands and lakes. The very localised contexts, like wells, are important in otherwise oxic areas. Most kinds of biological materials are preserved. Beetles and seeds are the most important in giving a range of environmental information at the site, local and sometimes climate scales, but more spectacular are the remains of whole woods, revealed as tree stumps at the base of peat cuttings or marine intertidal zones, and the remains of humans with their skin, nails, hair clothing and stomach contents. Timber is also important for dendrochronology (p. 139).
>
> ### Other burial environments
>
> Anoxic conditions in tar pits and the toxic environments of salt and metal mines slow decomposition and lead to the preservation of skin and wood. In the semi-arid American south-west, timber buildings are well preserved; in desert areas, e.g. Egypt, mummification aids preservation; and in Siberia, permafrost aids the preservation of the bodies, as in tombs in the Altai, and frozen mammoths.

PHYSICO-CHEMICAL CONDITIONS

In the first instance, depositional environments are characterised by pH, oxygen and temperature (Box 7.1). pH is acidity (sourness) or alkalinity (usually lime/calcium carbonate) on a scale from 1, acid, to 14, alkaline, with 7 being neutral. Soils with pH below *c*. 5.5 have low biological activity, and support few species of plants and animals, few worms and no moles. They are poor in nutrients and have a weak structure. Iron compounds and phosphates are susceptible to leaching. Polyphenol complexes derived from plants which thrive in conditions of low pH exacerbate these conditions. On the other hand, above pH 5.5, soils and sediments are increasingly rich in nutrients, having minerals in solution important for plant growth such as potassium, magnesium and aluminium; they have high biological activity, and support a diverse fauna and flora; they often have a strong crumb structure and are not so susceptible to erosion and leaching as are acid soils. Oxygen and temperature determine the rate of decomposition of organic matter by micro-organisms and biological activity generally; within limits these are higher at higher oxygen levels and temperatures.

Environmental landblocks

Landblocks which are defined by combinations of these conditions (Box 7.1) are one of the main frameworks of archaeology. Each has its distinctive soils, biota, hydrology and topography, and its own distinctive archaeology, environmental history and methodology. We see this, too, at the local scale as in the close juxtaposition of valley bottoms with their thick sequences and valley sides where there are thin soils and erosion (Figs 7.9 and 7.10). These differences, however, may have been less apparent in the past than they are today (Box 4.1). The implications of this are that the preservation conditions of today (Box 7.1) were not so widespread in the middle Holocene. This is seen in the increasingly impoverished pollen record in podsols as one goes back to when the soils were of circum-neutral pH, conditions unsuitable for pollen preservation, and in the paucity of molluscs from soils on areas which now support rendsinas. These environmental indicators are not uniformly present back into the past. We can talk about an archaeology of heathlands or wetlands today but this is not appropriate for the same areas for the middle Holocene.

MOVEMENT AND CHANGE

The next most critical division of depositional environments is between situations where lateral movement is substantial and those where it is minimal or absent (Table 7.1). We go from contexts in which there is extreme lateral movement, even in life (as with fast-flowing rivers), to those where the depositional assemblage is exactly co-located with the life range of the organisms (as with coral reefs).

Lateral movement

To start with a summary: lateral movement occurs during life and at different scales depending on the mobility of the organism or process, during death as with bird of prey kills, immediately after death as with the removal of carcass parts, during deposition, and after deposition as with the erosion of old sediments. No context contains an assemblage solely reflective of life on that spot, and clastic sediments, especially, contain materials of a variety of ages and origins (pp. 39–40).

DEPOSITION AND POST-DEPOSITION

Table 7.1. *Deposition and post-deposition of biological materials, soils and sediments*

Lateral movement during deposition

Post-deposition, disruption of deposits, e.g. by erosion of old river-bank deposits, soil erosion, human activities
    Long-distance, e.g. river-channel flow
    Short-distance, e.g. single spate events

Fine incremental deposition, e.g. overbank alluviation, blown sand; minimal disruption of biotic processes

No lateral movement during deposition

No downward movement
    Precipitates, e.g. tufa, lake marl
    Other autogenic deposits, e.g. peat, diatomaceous earth
    Some human occupation horizons and constructions

Downward movement
    Soils, as of pollen and snail shells (Fig. 7.8)

Turbation

This is *in situ* disturbance which can be co- or post-depositional

Bioturbation – Pigs, burrows, rootholes, tree-throw pits, total biological fabric, interstitial and burrowing faunas (Fig. 7.8)

Physical turbation – tillage, cryoturbation, ice-wedge formation, shrink-swell (especially clays)

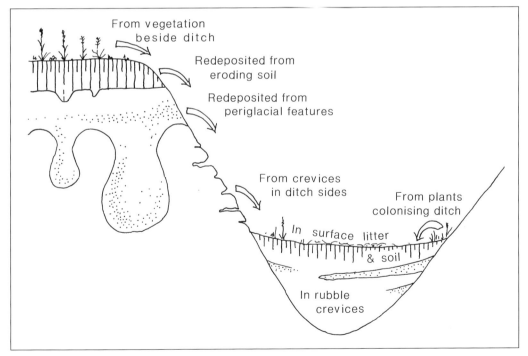

Figure 7.3 *Different origins of land snails in a ditch; all of these sources could contribute to the assemblage of a single layer.*

River deposits are among the most complex, deriving from a variety of environments and forming in a variety of conditions. Channel deposits are always mixed and floodplain deposits contain mixed assemblages from different seasons (pp. 160–1). Mixing also occurs when there is longer-term environmental change leading to the replacement of one community by another (Briggs *et al.* 1985). Briggs *et al.* (1990) relate death assemblage composition in terms of distance from life communities to particular floodplain units in the modern braided system of the River Lech in Austria. Modelling the movement of different groups of materials in rivers through experiment and observation helps to explain their patterning in fossil deposits (Hanson 1980). Stone artefacts and mammal bones are often associated (Wymer 1976; 1992), with the artefacts older than the bones because of different transport distances according to shape, size and density, while both groups can be older than the sand and gravel matrix.

Ditch fills are less mixed, although there can still be a variety of origins of materials and some of these can be from the re-deposition of old soils and sediments (Fig. 7.3). But many clastic sediments form gradually, allowing life to continue, and some plants and animals are specifically adapted to such conditions. Sand-dune grasses have extremely long roots which reach down to the water-table and allow the plants to survive as sand accumulates. Animals of floodplains have desiccation-resistant eggs or adults which can survive dry periods by burrowing into mud. Assemblages in these sorts of contexts can be virtually autochthonous.

Specific loci bring together general aspects of the surrounding environment, like the growth of woodland in a ditch in a generally wooded area, the build-up of manure in a midden in a generally arable one, or the deposition of sediment in a lake from a wider catchment (cf. p. 61). And certain loci not only attract but also make more visible general aspects of life which might otherwise be less obvious, such as the enhancement of declining soil fertility by differences of topography (pp. 34–5 and Fig. 3.5). These loci may thus contain two components, both reflecting the general environment, but one a passive incorporation of aspects of the ecosystem, the other an active development of it (which is a combination of the general environment and the specific properties of the depositional locus). However, only some of these loci preserve these features into the archaeological record whether in soils and sediments or in documents. For example, changes of solar radiation give varying climatic responses in different places as related to the distribution of land and sea, but only certain areas like ocean bottoms and ice-sheets will preserve long and detailed sequences of these. Or, an area may see considerable documentation in respect of land transactions, property disputes and inheritance rights, but only if there is a suitable repository for this material, such as a rich monastic foundation, will it be preserved. However, the situation is more complicated because the depositional loci engender their own specific life assemblages.

Two examples will make this clear. In the fill of a prehistoric pit, we may find the following origins of materials:

1. Materials from soil and sediments (e.g. periglacial involutions) around the pit (Figs 7.3 and 7.4) (=post-deposition).

2. Materials from the wider environment, like pollen (=deposition).

Figure 7.4 Section of an Iron Age storage pit at Balksbury, Hampshire, showing a natural fill in the lower part up to the dark stone-free layer and an artificial fill with large stones in the upper part. Note the periglacial involutions in the chalk, from which material was incorporated into the pit during the natural infilling.

3. Small vertebrate skeletons from animals which have fallen in (=locationally biased death).

4. Plants and animals which lived in the pit, e.g. carnivorous snails, blue-green algae (=life).

5. Materials which relate to the primary function of the pit, e.g. grain, and to its secondary function, e.g. rubbish which itself can be primary or secondary (Figs II.2 and 7.5).

6. Ordered deposition of specific categories of materials as manifesting structuring principles within human society (Hill 1995). Previous use of the pit, for example, granary, failed granary, basket- or clay-lined, burnt or unburnt, and the stage in the life-history of the material (e.g. Figs II.2 and 7.5, and cf. Box 6.2) are relevant to what is deposited.

7. Even when completely infilled, the location of the pit is clear as a dip since sediments compress and organic materials decay. The pit may then be re-incorporated (post-deposited) into future lives, for example as a part of ancestral continuity or the legitimation of activities through the place (Hingley 1996), with a house being constructed over it.

ENVIRONMENTAL ARCHAEOLOGY

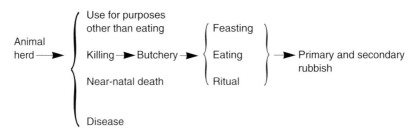

Figure 7.5 Different stages in the history of relationships of a bone relevant to its place of deposition (cf. Box 6.2 and Fig. II.1).

Towns and cities show this too. Pits, latrines and deep, complex stratigraphy constitute depositional environments typical of urban archaeology. Deposits contain pollen and insects which reflect the environment beyond the town. Towns draw together countryside resources, from a range of production strategies, each of which may be reflected on only one particular rural site (Addyman 1982). They have their own distinctive human environment of dense population, high intensity of social life, possibilities for mass organisation and the pressures of loneliness; and, partly because of these features, they draw in and intensify social phenomena which are present in society as a whole, rural and urban (Saunders 1985).

Minimal or no lateral movement

Where there is minimal or no lateral movement, the soil or sediment reflects stable conditions, and small-scale lateral variation is well preserved (Fig. 7.6). After sedimentation, organisms colonise as macro-fauna (worms, molluscs and crustaceans), meiofauna consisting of minute organisms just visible to the naked eye which live interstitially in sand (both groups usually within less than a metre of the surface), and microscopic organisms which extend down many metres (Gray 1981). Worms and land snails burrow into colluvial deposits and soils (Carter 1990a; 1990b) (Fig. 7.7). Rock rubble is particularly susceptible to colonisation by land molluscs and this occurs in cave deposits and the primary fills of ditches (Fig. 7.3) (Evans and Jones 1973), sometimes to several metres. Fortunately the assemblages have distinctive species, often carnivorous and subterranean.

Soils epitomise the problems of 'deposition' and 'post-deposition'. With reference to the substratum, all soil processes are post-depositional (Courty *et al.* 1989), while soil itself is a succession of depositional and post-depositional processes (see Fig. 10.5). Pollen, seeds and shells are moved down by biological activity and water percolation, so that contemporaneity of matrix and inclusions is not preserved. Earthworm mixing is one of the main processes, but mites, other fauna and small roots are responsible too, the processes resulting in a uniform 'total biological fabric' (Courty *et al.* 1989, 142). Some earthworm species cast on the surface, thus causing the downward movement of stones, archaeological materials, seeds and snail shells which are too big to pass through their guts. This forms a stone-free zone at the top of the profile and a zone of larger fragments of a variety of ages lower down (Fig. 7.8) which, because destruction is selective, is made up of components such as the tougher pollen grains and shells. There are higher concentrations of some

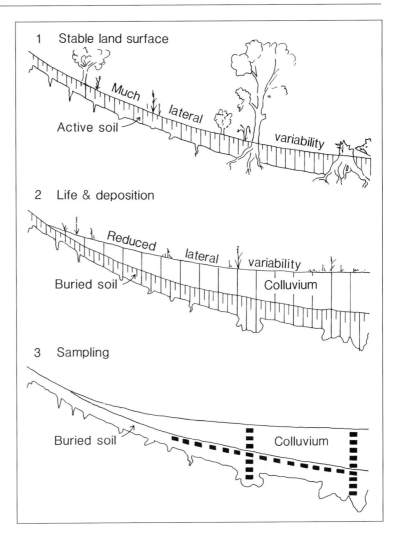

Figure 7.6 Changes from 1, a stable land surface, to 2, an unstable land surface with colluviation under arable; the sampling loci and intervals, 3, reflect the fine spatial variation and autochthonous assemblages in the buried soil surface and the reduced spatial variation and allochthonous assemblages in the colluvium.

materials at the surface, a useful criterion in the identification of ancient surfaces (Fig. 7.8). On a macroscopic scale, various processes collectively known as *litho-pedoturbation* (Table 7.1, above) disrupt soils and sediments. The snuffling of pigs, tillage and shrink-swell are uniform in their effects, often resulting in stones being worked up to the surface. Shrink-swell occurs through freeze-thaw and wetting-and-drying in clay soils, the latter especially in vertisols. Cracking (caused by frost or drying), cryoturbation, the effects of burrowing animals, and tree-root disturbance can disrupt the strata and their contained biology, and are usually visible as features like ice-wedge casts and burrows. Tree-throw creates hollows whose infill provides information about earlier soil history (Fig. 7.6). Even buried soils are not immune from change. There is compaction, reduction of organic matter through continuing biological activity (Bell *et al.* 1996), leaching of soluble materials, and downward movement and deposition of materials into the soil, all after burial.

ENVIRONMENTAL ARCHAEOLOGY

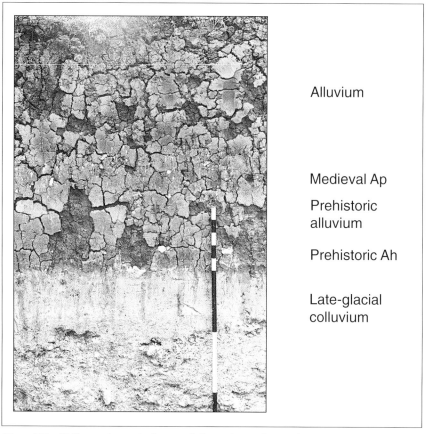

Figure 7.7 Valley section at Avebury, Wiltshire. The earthworm burrows in the Late-glacial colluvium contain cereal grains from the medieval ploughsoil (Ap), as shown by radiocarbon dating of individual grains (Evans et al. 1993).

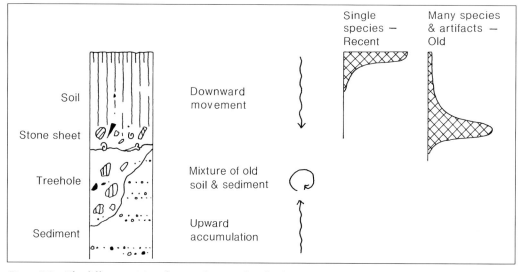

Figure 7.8 The different origins of materials in a soil and sediment.

In autogenic sediments – precipitates, biogenic sediments and some archaeological deposits like occupation horizons – where the matrix is formed *in situ*, artefacts, plants and animals stay where they are deposited by the processes of life. In peat, coral and algal reefs, and tufa, the bodies and skeletons of organisms make up the bulk of the sediment, although other organisms can burrow into the sediment and disrupt it. Even so, these deposits provide the least disturbed examples of past life.

POST-DEPOSITION TRANSFORMATION OF LAND AND ARCHAEOLOGY

Post-depositional processes transform the surviving data in relation to the nature of the archaeology (Figs 7.1 and 7.2) and land-use, erosion and sedimentation. However, it is often the case that post-depositional land-use is located in areas which saw the same kind of land-use in the past, as with the continuation of areas of woodland, cultivation or settlement through from prehistory, so that the land zonations we see today may reflect those of the past. Even so, successive periods of sedimentation and human occupation create zones of differences between life and the archaeological record, with greater surface diversity of sites and artefacts on eroding slopes, where they are conflated in thin soils, than on valley floors, where they are spaced out by sedimentation (Fig. 7.9).

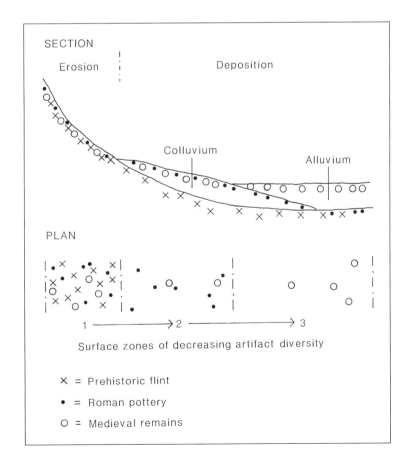

Figure 7.9 Generalised slope showing decreasing surface artefact diversity from slope to valley bottom.

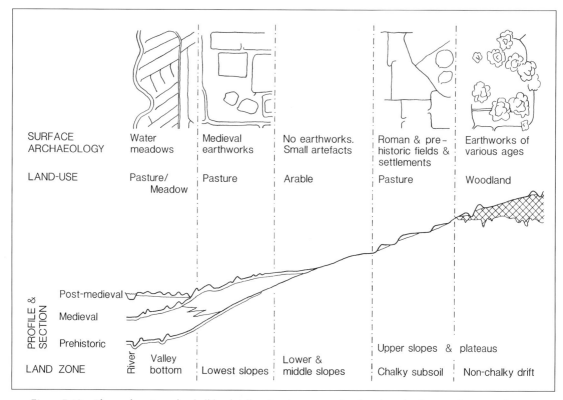

Figure 7.10 Plan and section of a chalkland valley showing zones of archaeology, land-use and preservation (after Evans et al. 1993).

In chalkland valleys in southern Britain (Fig. 7.10) we can identify five zones (Evans *et al.* 1993). On the plateaux, preservation of earthworks from the prehistoric period onwards is related to the distribution of clay soils, woodland and distance from later settlements. A similar situation obtains on the upper slopes, where preservation is again related to low-intensity land-use, especially sheep farming on downland. There is then a strip several hundred metres wide where few earthworks survive between the plateau/upper slopes and the lower slopes, which relates to intensive cultivation in the Middle Ages and later. Just above the valley floor are earthworks of medieval villages, while on the valley floor itself are watermeadows of the post-medieval period. Earthworks of the last two zones are preserved because of the absence of present-day cultivation; they often overlie colluvium which itself conceals earlier land surfaces of the Roman and prehistoric periods.

A transect from high mountain to coast in western Britain illustrates different types of preservation or destruction to which are related particular categories of archaeological visibility or concealment (Fig. 4.3; Table 7.2).

Table 7.2. *Zones of archaeology and environment in western Britain on an altitudinal catena from high upland plateau above the tree-line to the coast (Caseldine 1990)*

Region	Environment of preservation/destruction	Archaeology	Environmental data
High upland	Preservation in and under blanket peat	Hidden	Long sequences Peat pollen
Upland plateau	Preservation in areas unenclosed by historic-age farming	Visible on the surface	Short sequences in buried soils Soil pollen
Lowland	Destruction in areas of enclosure and settlement, except in ditches and under extant banks	Hidden or as cropmarks or surface artefacts	Patchy
Coastal strip	Preservation in areas unenclosed by historic-age farming	Visible on the surface	Short sequences in buried soils Soil pollen and molluscs
Large estuaries	Preservation under deep clays and peats	Hidden	Long sequences Pollen, molluscs and submerged tree stumps

Zones of destruction and preservation

Most past human activities are destroyed by weathering, agriculture, settlement and, in Mediterranean and semi-arid regions, erosion. Destruction can be total, as by quarries and river erosion, or partial, as by cultivation. However, the same processes can also lead to discovery (Stevenson 1975). Sites are revealed in gravel pits and through ploughing, which reveals surface artefacts, buried features like cist graves, and cropmarks or soilmarks. Erosion in semi-arid and tropical areas is also important, as with the discovery of early hominid bones and artefacts in east Africa.

Preservation falls into two categories. *Good spatial resolution*, where relationships are visible over wide areas, is seen in upstanding earthworks, cropmarks and soilmarks, and surface distributions of small artefacts and bones. The best conditions are where there is light ploughing, giving information about earthworks and small artefacts. However, the time/depth resolution is poor, with only short sequences in soils and local contexts like ditches. *Good vertical resolution* occurs where there has been burial of a land surface by later deposits like alluvium and blanket peat, often themselves with a sequence of land surfaces, and in urban sites and others with deep stratigraphy like tells and middens; but in these situations, because of practical difficulties, spatial resolution, although present, is difficult to study.

Representativeness of Preservation

Preservation of environmental and human activities is not a direct reflection of those activities.

1. Human activities often occur where preservation – either by deposition (lake edge) or abandonment (marginal land) – is likely; but other activities away from such areas are destroyed.

2. A preservation locus is one of many in the spatio-temporal continuum of life, and it is likely to be peculiar to a specific life activity (pp. 24–5), whether for people or other organisms.

3. The physico-chemical nature of the deposited materials makes preservation selective, with most contexts being devoid of all the important environmental indicators and only the more extreme ones containing them.

4. Soils and sediments where there is no lateral movement in their formation often reflect the regional situation, and lateral variation is well preserved. However, sediments formed by lateral movement are atypical both in the general sense and in that sediment sumps draw in materials during life and sedimentation, masking or distorting, but sometimes enhancing, the regional picture.

5. There is a positive relationship between buried ecosystems, whether under earthworks or sediments, and the environment that buries them (pp. 76–7).

6. At the landblock scale, destruction and preservation are related to land-use, erosion, sedimentation and soil processes (Table 7.2 and Fig. 7.10), although these may be co-locational with the same earlier land-use.

Part Three

PROCEDURES AND METHODS

We now turn to the procedures and methods of data gathering, especially their application to wider research questions, and the different kinds and scales of information they yield. Sometimes the recording of data is seen as an end in itself, or a sampling opportunity is seized on without a research strategy. An archaeological excavation leads to the discovery of datable deposits containing plant and animal remains and there is pressure to sample these for their own sake. But in our study of a crannog (chapter 5), if we looked only at the plant macrofossils from the occupation we would not be doing research. We would be ignoring other aspects of the operational environment around the site represented by animal remains, we would not be considering the wider region in which the crannog was set (except through resources brought onto the site), we would not be setting human occupation in the longer-term environment before the site was occupied and after it was abandoned, and we would not be considering the relationship between these data and those from other crannogs or dryland sites of the same age, or in other regions. All investigations must be set in a framework of research.

At the same time, projects need their own research agenda and should not be driven exclusively by questions posed from cultural or social archaeology. Ecosystems are at least as complex as social systems and merit investigation in ways and at scales which are not always appropriate to studies of past societies.

In a way, although this part of the book is about data collection and study, it is also the most critical. Sampling cannot be done without research strategies, while the processes of sampling, analysis and interpretation go on hand-in-hand. Each stage of analysis and interpretation leads to a renewed assessment of research directions and a concomitant renewal of sampling strategies. The sequence 'research strategy–sampling–analysis–interpretation' is iterative and cumulative (cf. pp. xi–xii).

Eight

RESEARCH DESIGNS AND SAMPLING STRATEGIES

THE GLOBAL VIEW

Research designs must have a global perspective, not necessarily in terms of geography but certainly in terms of fundamentals. The big aim in our work, which we have reviewed in a historical perspective in chapter 1, is to elucidate the relationship between environment and people – the human niche. For some, the study of past environments does not exist as a research discipline, rather as a set of investigative techniques. The biophysical environment limits the diversity and history of human life but is otherwise subordinated to sociality. For others, environment is so bound up with human activities as the human niche that its study is inseparable from archaeology. A moderate view would allow a role to both the biophysical and social environments, with the one buffering or developing the other, and with innate biological and psychological influences as a part of the equation. Archaeology is especially well placed to tackle these matters because it spans a great diversity of cultures and social groupings, and it embraces the evolution of humans where the key ingredients of sociality, material culture and language emerged.

SPATIAL AREAS OF RESEARCH

In our research we need to identify the different qualities and influences of area and time which occur in all ecosystems. Indeed, every scrap of data, however small and apparently atemporal, has within it information about a wider area and a wider past (pp. 61–2). Looking at this in reverse, these wider areas can be seen to be made up of small points of temporo-spatial data. A chip of wallplaster can give regional information in terms of the location and distances of quarry sites in a region as much as it can about the room from which it came, while a pollen sample from the middle of a 3,000 ha bog is just as capable of yielding information about the environment of the sampling spot – which may in fact be associated with some sort of human activity – as about the regional arboreal vegetation. Thinking like this gets us away from the assumption that on-site sampling explores the small-scale environment of the site whilst off-site sampling explores the environment of a wider area. In fact in chapter 15 we discuss an example – northern England in the Roman period – where the reverse is more nearly the case.

The scale of investigation is related to scales of human activity (Fig. 7.2), and not all of these are within the resolution of the archaeological record. Activities are diverse, even for a single village or region. The variety of scales and groupings to which ecosystem diversity and change relate – individual, community, biogeographical or evolutionary, and in the case of human societies, bureaucracies, religion or state – mean that we must keep flexible our scales of research. The smaller units – even the individual – must not be ignored, not just because these are what a region comprises, but because they may be relevant to the functionings of communities and to their diversity and change. Also, any one sphere of human activity can vary not only in area but also in the clarity of its boundaries through time (Cherry *et al.* 1978, 20), thus requiring a reassessment of the research area for different periods. So establishing the size of research units – individual, community, regional – and setting up a research design are not separate or fixed; each is reassessed against the other as research proceeds.

How not to define areas of research

It is often easier to say how areas of research should not be defined rather than how they should. Generally we should not use present-day archaeological, environmental or preservational partitioning of land as the basis, since these may be quite atypical, may pre-empt relationships, and may hinder inter-areal comparisons.

Areas based on concentrations of single monument types or rich collections of documents are a poor basis for research since they reflect only one of several activities, and that probably quite atypical. Particular types of artefact may transcend political or tribal boundaries (Hodder 1982b) or reflect only one of a diversity of activities in the life of a community. A study based on monuments samples only the modified environment, only one activity, an area that may have been special before the monuments were built and may be a context that allows a certain type of preservation. The wider operational and geographic environments remain unsampled.

Areas based on physical geography and landblocks of a particular environment – valley catchments, uplands and swamplands – are often taken as units of study. They are, however, unsuitable as frameworks for research because past human activities may have been unrelated to their boundaries. In the upper reaches of river valleys (Fig. 8.1), settlement is likely to extend beyond watersheds and even have separate lands in another valley (1) or it may be located on the watersheds and plateaux if the upland ecotones are sought after (5); in the middle parts, settlements may well reach precisely from the river to the watersheds (2), and this would be typical of farming villages; in the estuary and delta region, the valley may be so wide that several separate settlement systems are incorporated within it (3, 4, 6 and 7). Archipelagos and peninsulas are equally hazardous as areas of research, because estuaries and the sea may have provided stronger intra-regional links in the past than they do today. The clarity of present-day ecosystems may not have been so sharp in the past (Box 4.1), so one cannot refer to, for example, heathland archaeology for every period in the past because it was not always in heathland.

Equally, good preservational environments, often spatially localised, are a powerful draw, but such reactivity is a poor basis for research. For example, there is an attraction to blanket peat, the archaeology beneath it (Fig. 8.3) and its association with buried soils,

ENVIRONMENTAL ARCHAEOLOGY

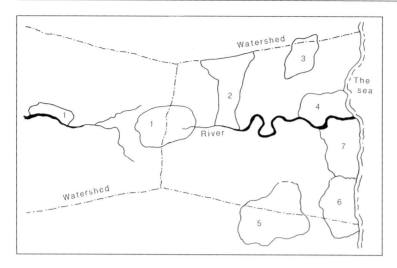

Figure 8.1 A river valley is a poor unit of regional research as seen in different distributions of settlement units (see p. 95).

pollen-bearing deposits and charcoal (Whittington 1983). But there must be investigations in the wider contexts of adjacent peat and lake basins and areas of similar topography but with no peat (Fig. 8.2). Obviously in the latter there are problems with preservation, but this only serves to demonstrate the atypicality of studying one particular context. Likewise, caves are physically and atmospherically closely circumscribed units with a strong draw for animals and humans in the past and an equally strong draw for archaeologists and geologists today. There is a high predictability that cave excavation will be profitable in archaeological or palaeontological terms or both, and yet the activities of animals and humans in caves are often specialised and only one of several that take place in the wider ambience of the surrounding lands. This point has been made, for example, by White (1983) with regard to the asymmetry of settlement data for the Palaeolithic in south-west France (although see Box 14.3).

There is a cumulation of biases in that sites in life which are a strong draw to human activity, are likely to be those that are most visible in death, those which have the best conditions of preservation, and those most noticed and studied by archaeologists (cf. Fig. 12.13) – all to the detriment of the study of the full operational environment of past humans.

Defining areas of research

Areas delimited by national or global grids are probably the best to start with, especially if they are large enough to encompass several major units of archaeology, topography and preservation. This enables us to test variations in human behaviour against environment within and between them (Gamble 1986, 306). Areas should be defined in terms of the archaeological record only with reference to a variety of archaeological categories, visibilities and documentation. If a particular land and preservation type is the focus of study, e.g. an arid plateau, then at least one other, e.g. a major river valley, should be

Figure 8.2   Eroded hard limestone pavement, The Burren, Co. Clare, Ireland. This and the next plate show the different effects of later Holocene change as related to geology and drainage.

Figure 8.3   Blanket peat accumulations on sandstone, north Co. Mayo, Ireland, showing a pre-peat wall (p. 158).

ENVIRONMENTAL ARCHAEOLOGY

Box 8.1 South Nesting, Shetland: scales of sampling

The archaeological investigation of a multi-period landscape at South Nesting, Shetland, in 1991–4 presented the challenge of sampling different scales of time and space (Dockrill *et al.* 1998). One of the attractions of the area was the way in which geology, topography and prehistoric and recent settlement showed a series of differing influences and associations within the modern landscape.

Defining the study area

Defining the study area was not simple. Much of its limit was the modern coastline, but it is not clear that this was the coastline, say, 3,000 years ago. There are peat deposits buried below intertidal deposits, showing that a rise in sea-level occurred in the Holocene, with implications for the location and availability of resources (pp. 165–6), and for the drainage and geochemistry of groundwater. Other parts of the study area were defined with respect to present-day vegetation and land-use. This is obviously arbitrary, particularly as these parameters were affected by nineteenth- and twentieth-century crofting. However, one of the questions addressed by the project was the visibility of the surviving archaeological record with regard to modern land-use patterns.

Defining 'sites'

Fieldwalking of the whole study area enabled the identification of concentrations of structures and artefacts. Geophysical and small-scale topographical survey around these further tightened the definition of the individual 'sites', allowing limited excavation to be planned as informed judgemental

The main land units are: a = unenclosed land mainly above 40 m OD, largely blanket peat; b = unenclosed land mainly above 30 m OD, largely ranker and heath soils; c = enclosed land, mainly at or below 20 m OD, of nineteenth- and twentieth-century settlement and crofting; d = shallow marine inlet with peat below intertidal deposits. The research design encompasses each of these, whilst allowing that prehistoric archaeology might show a distribution similar to recent settlement, even though its survival is actually complementary to it.

South Nesting, Shetland: scales of sampling

sampling. Excavation trenches deliberately sampled both structures and the gaps in between them, but were none the less concentrated in 'monument', rather than 'non-monument' areas.

Intra-site sampling

Deposits encountered during excavation were sampled at diverse scales for diverse reasons. At site SN177, a peat core was retrieved from below a 'burnt mound' in order to examine the vegetation of the area when the burnt mound was constructed. However, the core was extended to nearly one metre below the base of the mound in order to gain the vegetational history leading up to the mound construction. The pre-mound environment could then be seen as one stage in a continuum of vegetational change, not as an isolated snap-shot. The core also sampled the regional environment in the form of extra-local pollen in the peat, so giving information at a number of spatial scales.

At site SN229, a palaeosol formed the B horizon of the modern soil and appeared to be a cultivated prehistoric soil. The soil was sampled by Kubiena boxes for micromorphology, to analyse inputs to it, its use, and its subsequent pedogenesis. Results gave information at the scale of the blocks of soil retrieved and examined, and at that of changing regional land-use and settlement. The soil was also sampled by several well-separated trenches to examine its surviving extent and spatial variation in thickness and appearance. This was essential to give a frame of reference to the micromorphology samples.

In this one field project, then, sampling decisions were taken at all levels from the initial selection of the survey area to the precise placement of Kubiena boxes, and the sub-sampling of a peat core for pollen analysis. The extra-local component of the pollen from each 1 cm^3 pollen sample gave information on a scale much larger than the roughly 12 km^2 of the study area.

Section through the edge of a prehistoric field. The profile to the right of the boundary shows the modern soil developed on an anthrosol, with ploughmarks on the subsoil surface; the stones in the shallower profile to the left were probably cleared from the cultivated area and dumped there deliberately.

compared. Where a town or city is the main area of study, this should parallel a relevant rural area, and comparisons should be made with another town. This, in theory at least, separates characteristics of the past life or preservation which are specific to one area and those which are more widely relevant.

Regional research projects involving surface surveys and settlement and environmental histories have been undertaken for areas of well over 100,000 km². This is too large an area to study in minute detail so the regions need to be sampled (Fig. 8.4).

Detailed study can be done by selecting areas of different land types (lowland and upland), archaeology and documentation (dense and weak) and enclosing them in arbitrary rectangles, as Woodward (1991) did in his Dorset Ridgeway study. Non-monument *vs.* monument areas is an obvious contrast for an initial division of a region (Figs 8.4 and 8.5). So, in the same way, is the contrast between major soil types, which will strongly influence the survival and visibility of archaeology (see Fig. 9.4) and may also reflect the distribution and intensity of past human activities (Box 8.1). Bintliff (1997), in his Boeotia survey, adopted a different approach, with total surface survey of four adjacent communes (or parishes) designed to identify activities associated with individual farming settlements. Davies and Astill (1994), in their East Brittany survey, although taking the same number of adjacent communes within their region, looked at alternate 2 km-wide transects and in these were unable to do a total survey because of the amount of land under pasture. The Boeotia and East Brittany surveys both used geographical and historical regions for the basis of their main areas, but Bintliff (1997) took things a lot further in making intra-regional comparisons of both survey and settlement-history data throughout Greece and beyond (pp. 211–12).

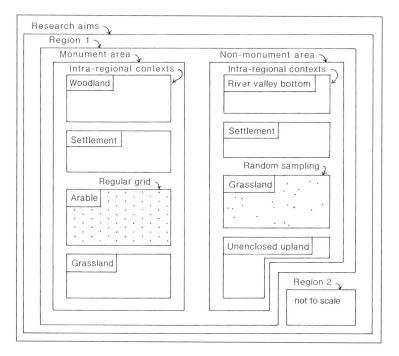

Figure 8.4 A planning aid to sampling strategy for two regions (see p. 100; also Fig. 9.1, p. 107). Note the different categories of sampling, although there is no significance in their use here in the different land types.

RESEARCH DESIGNS AND SAMPLING STRATEGIES

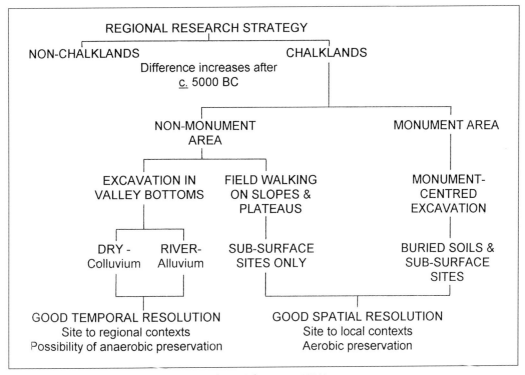

Figure 8.5 Sampling strategy for a chalkland area (after Evans 1993).

INTRA-AREAL SAMPLING

The physical environment

Initially, the aim is to detail the past physical environment, its distributions and densities (chapters 3 and 4), the influences of different spatial scales and qualities of environment, and temporal sequences. Present-day environment (vegetation, climate, land-use), archaeology (including documentary and toponymic data) and resources for environmental investigations are mapped (chapters 9 and 10). Contexts are then selected for sampling not only for detailing past environments but in order to identify the influences and modifications that took place during death (chapter 6) and preservation (chapter 7).

It should be clear by now that every context contains information of different scales and qualities of space and time, not least of which are those of the context itself (e.g. Chapter 5). To summarise, these can relate to: the general environment – climate, biota and soils – and the different reflections of these in gradients (e.g. Figs 3.1, 3.5 and 4.2); the enhancement of general aspects of life from beyond the context; the buffering or enhancement of these by specific features of the context, creating smaller spatial contexts (e.g. Figs 4.5 and 4.7); transportation by humans, biological dispersal and sedimentation; and modifications by death (chapter 6), preservation and post-deposition (chapter 7).

In order to try and separate these different components, three main sampling strategies are used.

1. As great a variety as possible of archaeological, land-use and preservational contexts is sampled (Figs 8.4 and 8.5), and by various methods (chapter 9). Modern settlements, boundaries, even woodland may have been continuous for long periods, possibly into prehistory, and must be included. There is a tendency to focus on areas of 'good' potential which are often in marginal land and to shy away from more populated and used areas (Davies and Astill 1994); only urban archaeology has tackled this systematically. Other land-use types are selected for their different destructional or preservational signals (pp. 80–1) and the different methods applicable to them (Fig. 9.4).

2. Different types of spatial variation are identified with reference to basin size and type, soils *vs*. sediments, and biological materials. Basin size is relevant to the area from which materials derive, while location in the basin is relevant to separating influences of regional, local and sump-specific characteristics (Box 6.1). A spatial picture can be built up from multiple samples. Sediments are selected for their depth, with fine temporal resolution and unequivocal sequencing, while their spatial significance is assessed from their location, their overall distribution and their intrinsic properties. Lateral sampling need not be close in clastic sediments because they are composed of materials from a wide area (Fig. 7.6). Soils provide specific detail on a site scale while at the same time being more typical than sediments of the regional environment. Lateral sampling of soil surfaces should be close spatially in order to identify this detail (Fig. 7.6), and may need to cover a wide area if there is local variation. The picture is expanded by biological materials, with scale of information relating to their mobility in life (pp. 68–9 and Table 5.2); for example, sampling for pollen need not be as close laterally as sampling for snails. Sampling for these data requires a consideration of their inherent biology and ecology, of the human activity which may have been involved in their deposition, and then the depositional circumstances.

3. The location and sampling of long, complete sequences. These will only be available from a few loci. Often they seem irrelevant to specific research aims, especially if these are focused on a single, narrow time horizon, but they are vital in providing an indication of the variety of time influences on that horizon, and in giving a context to the narrow time horizon.

Questions

All fieldwork turns up materials which have arrived at the sampling spot from somewhere else, and the question now is: 'From where?' It is especially relevant to the origin of materials moved around by human activity. To go back to our crannog (chapter 5), the occupation level has cattle bones in it, showing that people were discarding bones at the crannog and probably that they were eating meat there too. But they also tell us that cattle farming was going on around the lake. So we need to ask where the animals were kept, what sort of environment were they in and how were they fed during winter. One approach is through isotopic and other chemical analyses of the bones themselves, which give us information about the diet of the animals and hence where they might have fed

(p. 148). Another is examining the land around the lake for field systems and enclosures where the animals might have been kept (p. 157 and Fig. 12.7). These in turn can incorporate contexts like colluvium, buried soils under banks, and distributions of artefacts and geochemical materials (chapter 9), which can indicate not only the way the fields were used but also aspects of the wider environment, especially the land-use of the area before the fields were constructed. A third approach would be to do pollen analyses from the lake sediments and peat on the hillslopes and plateaux to identify the spatial distribution of plant associations. If enough sampling spots were used, the distribution of pasture, arable and woodland in the area might be determined. Pollen analysis might also tell us something about lake levels, important in relation to rebuilding phases on the crannog, or local activities at the sampling spots such as peat cutting or burning, and other uses of the wetlands around the lake.

In all this we are going from a small location of bones at the crannog to a wider distribution of fields around the lake and then on to further questions.

Why?

Why was a particular area chosen to graze cattle? Why did the wallplaster of a house come from a particular locality and not another one? Why was a trackway put across this wetland area and not that one? This is moving on to interpretation, but we cannot avoid it even at the sampling stage because it is relevant to how we proceed. Once we have located our fields in which cattle might have been kept, we need to consider what other possibilities for their location there could have been, why they were not so located, and what else might have been going on in those areas. This is where a broad archaeological view is needed because the answers may lie in competition with other, dryland, settlements contemporary with the crannog, but having different land-use systems, or with wider social relationships concerning the use of particular areas of land (cf. pp. 15–16; p. 56; Box 14.2). Ultimately, we are asking how the past human population thought it appropriate to use a particular area of land, a question which in part addresses functional decisions and in part ideational ones.

SMALL-SCALE STRATEGIES

An occupation site can be seen as a focus of human-made habitats, possibly of many different kinds (hut interiors, refuse pits, disturbed soil surfaces in arable plots), interdigitating with, and showing different forms of transition to, a range of natural habitats which may show some degree of modification by human activity (deciduous woodland, alluvial mudflats, sandy littoral). There is a mosaic of habitats showing more or less intense human modification, varying in resilience and stability and changing at different rates through different successional processes. It is clear that the location of our sample with respect to the location of human occupation will largely determine the data obtained and the questions which can be addressed through those data.

A nice example of small-scale sampling comes from Beverley (Yorkshire) in eastern England (Evans and Tomlinson 1992), where samples were taken through medieval refuse deposits that overlay an earlier land surface. Samples from the upper few centimetres of the buried soil at the bottom of the sequence yielded botanical remains indicating plant species associations typical of damp alluvial soils. These data gave information about certain attributes of the environment on a scale, probably, of a few tens of square metres, and with appreciable time-depth. The palaeosol was overlain by a black humic mass consisting largely of vegetative fragments from monocotyledonous plants, which was interpreted on the basis of its morphology and contents as a mass of horse droppings. Assuming the horses to have ranged fairly widely when grazing, this deposit sampled plant communities on a scale of several hectares, though the proximity of the life range of the plants to the point of deposition of the droppings is not known. In the top of the 'dung heap' deposit was a clutch of unhatched snake eggs. These gave very small-scale, time-specific information, namely that the top of the deposit was warm enough to be attractive as a place of incubation, and that subsequent deposition probably ensued before the eggs could hatch.

METHODS

Our choice of techniques depends on what we need to know, and therefore on our research and sampling strategy. There are two important concepts: the data we can see and touch, and the inferences that can be made from them, referred to respectively as

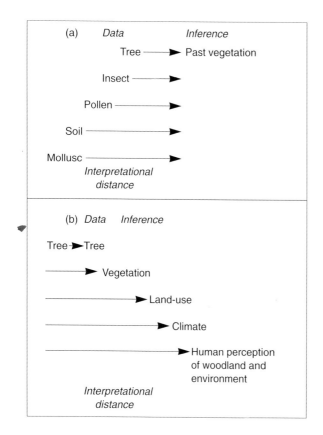

Figure 8.6 Distance between data and inference for (a) the same inference, (b) the same data.

actualistic and *proxy*. Actualistic data are a collection of beetle elytra (wing cases) in a well, the inferred proxy conditions the vegetation and climate they reflect. There is a scale of certainty, from the beetle elytra (actualistic) implying a breeding beetle population (proxy, certain) to that population implying a stated environment and climate (proxy, less certain). The further we go away from the raw data and the more we ask questions about the abstract, usually the more useful questions, the less secure we are about the answers.

The distance between data and interpretation varies depending on the nature of the data and what we want to know. To take a single aspect of the environment like vegetation (Fig. 8.6a), there is an increasing distance between it and the data as we go from preserved tree stumps, through pollen and invertebrate animals, to soils. This is related to a combination of the data themselves, mobility of the materials in life and their taphonomy. Seen through a single category of data (Fig. 8.6b), there is increasing distance between timber remains and trees, through vegetation, human land-use, climate, to past human perceptions. The tops of these diagrams are actualistic data – the tree in the fossil record indicates a tree in the past – while at the bottom are proxy data, where we are inferring a particular type of vegetation from a particular type of mollusc. All the data are actualistic for something and proxy for another; it just depends on what aspects of the past we are trying to interpret. Past climate is particularly intangible and can only be inferred from proxy data.

Interpretation of the different stages from raw data to environment is discussed in Part IV, but since it involves another category of techniques, it is usefully introduced here. The techniques in question are called *possibilistic*. They are fed into the interpretative procedure between the raw data and the environmental factors – proximal and distal – to which they relate (Figs IV.1 and 14.1), indicating possibilities rather than certainties.

NINE

METHODS OF EXAMINING ARCHAEOLOGICAL DISTRIBUTIONS

The distribution of field archaeology is a part of the past environment, partly as a contribution to its physical structure and partly as a reflection of human activities. The building of a hogan or villa affected people's ideas and use of the land as much as woodland clearance or coastal flooding. So the methods described in this chapter, although essentially developed to map archaeological data, are directly relevant to past environments (Aston 1985, with a useful bibliography; Brown 1987; Clark 1990). The use of any one technique depends on the scale and aims of the investigation, the contemporary environment and land-use, and the nature of the remains (Fig. 9.4). The techniques are presented in the framework of a regional research strategy (Fig. 9.1).

RECORDING SYSTEMS

Foremost, we need a system whereby we can record, analyse and interpret at the whole range of scales of human behaviour. *Geographical information systems (GIS)* provide a valuable computer-based technology which stores spatial data in map form, allowing comparisons to be made between distributions of archaeology and environment (Table 9.1) (Andresen *et al.* 1993; Lock and Stancic 1995; Maschener 1996). It also enables abstract aspects such as the views from particular sites in terms of other sites and topography, and the erosion potential of soils, to be incorporated (Gaffney *et al.* 1995). We can also feed in temporal information (Langran 1993), such as changes in the environment of an area, or the time it takes to get from one place to another using different means of transport and under different conditions of vegetation and terrain. Some of these aspects are the province of interpretation (e.g. Box 14.4), but we emphasise them here because of the value of GIS from the start of the recording process.

MODERN MAPS

At an early stage maps of geology, topography, soils, land-use and archaeology are essential if available, since our interpretation of surface archaeology is greatly hampered without them (Fig. 9.2). We need to identify the lithopedo-stratigraphy, areas of erosion *vs.* destruction, and deposition *vs.* preservation, and locate places where archaeology is likely to be close to the surface and places where it is likely to be buried (Table 7.2 and

METHODS OF EXAMINING ARCHAEOLOGICAL DISTRIBUTIONS

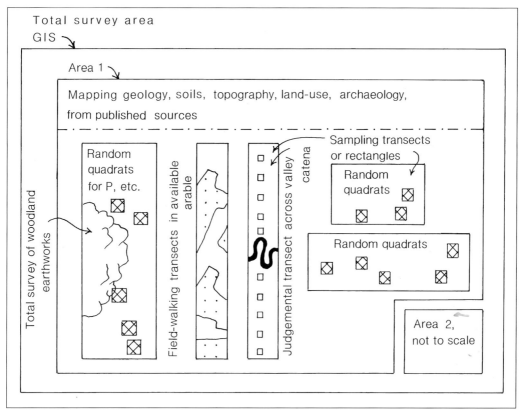

Figure 9.1 A strategy for sampling an area within a region (see Fig. 8.4, p. 100).

Table 9.1. *Various uses of GIS*

1. Inventory and display of conventional cartographic data, with overlays of different environmental parameters, and updating inventories as information is accrued.

2. Inventory and display of spatio-temporal attributes, as with the different age surfaces and landblocks of river valleys (Box 10.2).

3. Plotting and modelling environmental change, as in pollen assemblages from two or more phases; what has changed, and by how much?

4. Mapping space–time ratios, as with estimations of travelling distances under various conditions of terrain, vegetation and topography, and estimating rates of spatial change, as along a catena.

5. Mapping cognitive space, as with identifying views from particular areas or sites.

6. Modelling past and future change (Box 14.4).

7. Scheduling and forecasting when thresholds are reached and change is likely to be triggered.

8. Estimating how observed distributions might behave outside surveyed areas (Wheatley 1996).

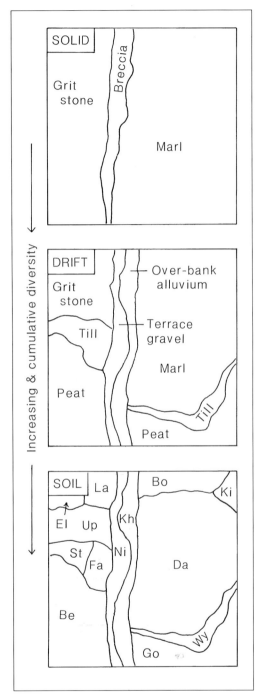

Figure 9.2 Solid and drift geology and soil maps for the same area; the abbreviations on the soil map refer to soil series.

Fig. 7.9). These features, too, are important sources of archaeological and environmental history, partly through their distributions and partly their contained materials.

Solid geology (usually pre-Pleistocene) aids in understanding topography, water sources and the origins of inorganic materials such as stone, clay and mineral ores. However, mapped units are often chronostratigraphic, and as such a single unit can vary in its hardness and chemistry. Other maps may show Pleistocene and Holocene unconsolidated deposits (the drift geology), which mask the solid geology and can be quite different, as with peat over free-draining limestone or chalky till on acid sandstone. Drift is relevant to fieldwalking because blank areas of artefacts may not reflect former inactivity areas if the drift is masking old surfaces (e.g. Fig. 7.9). Soil maps (p. 35) are also invaluable since they show variations in texture, drainage, pH, stoniness and depth which are excluded from drift maps where deposits are less than 0.4 m thick. Also, some soil properties are related to factors other than the underlying substratum, like vegetation and land-use (e.g. Davidson and Simpson 1984). Soil distributions are thus a cumulation of solid and drift geology and intrinsic properties of the soils themselves (Fig. 9.2). Land classification and land capability maps show various types of present-day land-use, such as grades of agricultural land, forestry and built-up areas. However, they can be misleading as an indication of past use, not just because of environmental change but because they build in extrinsic factors relating to present-day human use (Davidson 1980, 31). Archaeological distributions may be available from published sources and archives such as the Sites and Monuments

Records (SMRs) held in parts of the UK, mostly on a county basis and with county archaeology units or in local government offices. The archaeological data may be classified in terms of the quality of survival, and the quality of the original record.

Aerial Photography

Aerial photographs allow areas of ancient settlements, enclosure patterns and trackways to be seen in a topographical and vegetational setting (Wilson 1975; 1982). They also reveal ancient topography and other environmental features such as river courses, ice-wedge patterns and soil accumulations (Lowe and Walker 1997, 21–2). Coverage varies from the individual site to the regional scale, but it is at the local scale of, for example, the individual settlement and its hinterland that the most useful results may be obtained. Aerial photography, perhaps more than any other method (except fieldwalking, p. 114), has demonstrated the importance of areas between sites.

Aerial photography can be applied to ploughed land, where features show up as soil marks, to land under crops, where they may show as dark or light areas of differential ripening, and to pasture, where upstanding remains of the banks, mounds and ditches show as shadow marks. Built-up areas, alluviated river valley bottoms, clay soils and woodland, except for the latest episode of land-use, are less suitable, although it is surprising how much of an ancient pattern can show through (Aston 1985). How representative of former real distributions is the evidence of aerial photography? Some concentrations of sites may be related to the past importance of adjacent sites, such as the clusters of sites around Stonehenge in Wiltshire (England), although it may be, too, that these areas have been more photographed. Likewise, the concentration of sites on river gravels (e.g. Benson and Miles 1974) may not only show the suitability of those deposits for preserving cropmarks but also their suitability for settlement and cultivation in the first place. But, what was going on in the blank areas, further up the slope and on the river floodplain adjacent to these concentrations? They may have been devoid of settlement, at any rate of a kind that leaves cropmarks, or the evidence may have been destroyed by intensive ploughing (p. 90) or buried beneath alluvium. The blank areas need surveying by other methods if we are to get complete distributions.

Aerial photography may be enhanced by using infra-red imaging, and a wider range of imaging devices is available in orbiting satellites. Satellite-based devices give coverage of much larger areas than aerial photography (e.g. 185 km across for Landsat images), albeit at a lower resolution, and allow the filtering and processing of a wide range of light-wave bands. Airborne radar and seismic surveys have also been used successfully, for example to map landforms which are now underwater, and to map sediment interfaces across large areas (Lowe and Walker 1997, 22–3).

Historical Documents and Toponymy

These provide information about past environments for historic-age periods, although the data can be relevant to prehistory, for example where documents refer to prehistoric monuments, or where placenames are of prehistoric origin.

Maps

Maps often give the most direct evidence, and can be used in various ways. Like aerial photographs, they show spatial relationships of geographical and human-made features, which cannot be appreciated on the ground. Maps also show land-use sequences, as where a Roman road crosses fields (Fig. 9.3) or where open-field furlongs are subdivided. These data can be amplified by fieldwork and documentary evidence for the age of boundaries and buildings, allowing us to expose successively older environments, as has been done at Roystone Grange in the Pennines (Hodges 1991). Given a series of maps for one area, land-use history, such as the layout of boundaries, buildings, cultivated areas and woodland, can be followed over several hundred years (Rackham 1986). Maps can record ownership, tenancy, farm and field sizes, and land utilisation, so they have an indirect use as indicators of land value and productivity. Maps may also show the location of archaeological sites which have been destroyed, such as when a boundary makes a sharp angle around an ancient settlement.

Documents

These vary from estate accounts, which detail areas of land units, numbers of tenants and land-use, through medieval manorial accounts and surveys, which are often less direct in their information, to early medieval charters of land boundaries. In the British Isles, the earliest documents are boundary surveys of Anglo-Saxon estates, which are mostly of the ninth century AD onwards. They often give time-depth in their descriptions of prehistoric sites which were features of the early medieval landscape. In Ireland, similarly early documents record a range of landscape features, such as types of field boundaries in the seventh century AD (Reeves-Smyth and Hamond 1983). As with many documents, the eleventh-century Domesday Book gives environmental evidence of two kinds: direct, as in the distribution and extent of woodland, meadow, arable and pasture, mills, fisheries and salterns, and indirect in the extent, value and size of holdings which give information about productivity (Jones 1986). Medieval and later documents in Britain include hundred rolls, which deal with administration at the sub-county level, ecclesiastical cartularies which describe the distribution and activities of monastic granges, and documents like charters, court and account rolls, surveys and terriers covering smaller areas. At a regional or countrywide scale there are ecclesiastical, agrarian, enclosure and industrial histories and population statistics. Censuses go back to at least 1563 in England (Hodges 1991, 114), and there is much earlier information from the Classical World (cf. pp. 211–12).

A study of open fields (Hall 1982) shows how several types of data can be integrated. Archaeology indicates the pattern of furlongs and their lands, while maps of field names and lists of furlong names allow their allocation to specific furlongs on the ground. Fieldbooks and terriers which list individual owners and the location of their strips enable more detail to be added, as well as areas of ley grassland, meadows and woodland. In some cases, marginal land was taken into cultivation relatively late, as shown by the shape and context of the holdings, long tongues of land stretching into areas of poorer soils (on heavy clays or very acid sands), sometimes several adjacent from neighbouring parishes, and their field names which indicate woodland or heath. Later, these areas were often the first to have been put down to grass.

A word of warning about maps and documents. Their data are always difficult to interpret in environmental terms because of the influence of other factors. The turning over of arable to pasture, enclosure, settlement desertion, reduced rentals and falling populations may be interpreted in terms of, say, declining land productivity only if backed up directly by soil analytical data. Otherwise, a general trend to emparkment or sheep farming may have been the cause (Brown 1987, 115).

Toponymy

Placenames require specialised knowledge in their use because of their often multiplicities of meaning and the changes they have undergone. They are most useful for the periods and areas in which they were first widely adopted, as in the British Isles where British or Celtic names are mainly in the north and west, and those of Scandinavian and north Germanic origin (essentially English), established from the fifth to the tenth centuries AD, in the south and east. Placenames are topographical or habitative, and it is the former that are of most value to archaeology because they are often precisely descriptive, with different words for wetlands, hills and rivers indicating subtle properties of topography, settlement and land management (Gelling 1984). Along parts of the Rhine in Alsace, for example, different morphological and soil zones each have different colloquial, field and settlement names: *Auwald* (riverine forest), *Worth* (a river island), *Feld* (field), *Matt* (meadow) and *Ried* (reed) zones (Schirmer 1988).

Placenames are best used in conjunction with other evidence (cf. Hall 1982, above). For example, there is a relationship between soils and the chronology of settlement in the distribution of Scandinavian placenames in eastern England (Gelling 1978a); earlier names and larger settlements occur on better soils and areas most suited to agriculture, whereas later and smaller settlements occur in areas of apparently poorer resources, although in the latter, location of water may have been an over-riding influence. Present-day parishes sometimes correspond to Anglo-Saxon estates (p. 110) and these in turn may have been established in prehistory, as in parts of Berkshire where they are crossed by Roman roads (cf. Fig. 9.3). Only in heathland are parishes laid out with respect to roads, showing that this land was taken in at a late date; and 'there is a consistent contrast between the names of parishes which are bounded by the road, and those which ignore it' (Gelling 1978b, 125).

Two general points. Maps, documents and placenames proliferated in times of economic and social change and thus often relate to extraordinary periods of history. One can take a collection of monastic charters like the east Breton *Cartulaire de Redon* (Davies 1988) and use it as the basis of regional research (Davies and Astill 1994); but why this particular collection for this region and this period of history was so full and so well preserved has a bearing on our interpretation of the data and needs to be addressed. Also, the data for all these methods have a time-depth; they reflect change by copying, language shifts and even meaning. Some historians tend to see this as a nuisance (e.g. Alcock 1971), but it reflects the history of the sources just as do collections of disarticulated and fragmented bones (cf. Fig. II.2).

ENVIRONMENTAL ARCHAEOLOGY

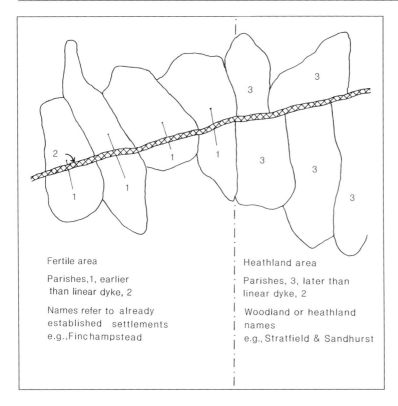

Fertile area

Parishes, 1, earlier than linear dyke, 2

Names refer to already established settlements e.g., Finchampstead

Heathland area

Parishes, 3, later than linear dyke, 2

Woodland or heathland names e.g., Stratfield & Sandhurst

Figure 9.3 Use of a variety of data in sorting out land chronology on the basis of land fertility, chronology of parish establishment as indicated by their relationship to a dyke, and placenames in relation to environment (loosely based on Gelling 1978b).

GROUND-LEVEL SURVEY

The succeeding techniques require close sampling, so further selection, now within the main transects or rectangles, is required (Fig. 9.1). Land-use and soil properties, especially depth, are crucial to the choice of methodology (Fig. 9.4).

Above-ground survey

This records slope changes and the distribution of stone – buildings, trackways, vermin traps and unenclosed land; in stony terrain, areas free of stone can indicate areas of cultivation (Fig. 4.6), useful in the absence of field boundaries. The method is slanted to use in areas which have seen little or no cultivation, with extensive 'official' mapping of upland earthworks; but this has been to the detriment of sites in woodland, and to modern settlements and boundaries, many of which may be of some antiquity. Many northern European field systems, villages and towns were established *de novo* in the Middle Ages or later, but many others overlie earlier settlements which go back to prehistory (Aston 1985; Johnson 1985). Field survey of upstanding remains has also been traditionally used on sites for which it is particularly suited for teasing out details of wall construction and chronology. Brick and stone buildings can be examined by infra-red thermal imaging to detect earlier phases of half-timbering (Fig. 9.5). But the

METHODS OF EXAMINING ARCHAEOLOGICAL DISTRIBUTIONS

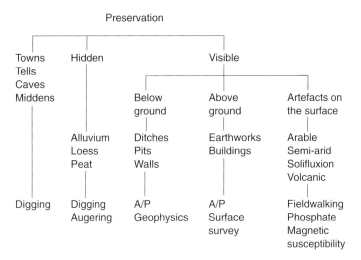

Figure 9.4 Methods of prospecting and survey according to conditions of preservation and archaeology (cf. Fig. 7.1).

site approach has been to the detriment of the areas between sites, and now it is standard practice to extend surveys more widely, especially with the help of aerial photography. The trend away from the site-centred approach has been aided in the last twenty years by the use of the electronic distance measurer (EDM) which has speeded up large-scale surveying, not least by allowing the downloading and printing out of data in the field.

Figure 9.5 The use of infra-red thermal imaging. Left: thermal image of a Saxon doorway, otherwise invisible; right: the doorway revealed by removal of the wall rendering (Photo: the Demaus Partnership).

Fieldwalking

Data on soil depth and history are vital to the interpretation of fieldwalking, geophysical and geochemical data (Fig. 7.9). The study of pottery, stone artefact (and building debris, charcoal and burnt stone) distributions on contemporary ground surfaces is used in areas of surface instability. In temperate climates, it is largely confined to arable land, but in Mediterranean, semi-arid and arid climates where there is much surface erosion, it is more widely applicable. Indeed, there is a gradient of increasing surface pottery densities from England, through the Mediterranean countries, to Arabia, which may be a function of increased soil erosion (Bintliff 1992). Precise conditions are important, and there can be differences in the archaeological signal between deep and shallow ploughing (e.g. Gardiner 1980). Fieldwalking is especially useful at the local scale where it focuses on activities that take place in hunter-gatherer daily/annual catchments or agricultural settlements. This level of analysis is often missed in the smaller and larger scales of, respectively, site and regional studies (Gaffney and Gaffney 1988; Schofield 1991). Analysis is most productive when combined with other data such as field boundaries and settlements, as in the case of Roman remains at Maddle Farm, Berkshire (Gaffney and Tingle 1985). Interpretation is difficult because of the various processes by which artefacts accumulate. The record of centuries or millennia can be conflated into almost zero thickness on erosion surfaces, while hillwash and alluvium mask artefacts in valleys (Fig. 7.9).

For pottery, sherd size and distribution density can indicate the stage in the pottery life-cycle (Fig. II.2, pp. 58–9) and the types of activity that go with it. For example: (a) Pottery associated with building debris (from fieldwalking) and buildings (aerial photography and excavation) is likely to represent settlement. (b) Dense concentrations of large sherds but with no building debris may represent discard at foci of non-settlement activity such as middens. (c) Wide low-density scatters of small sherds are likely to represent discard as manure in areas of cultivation. (d) Low-density or blank areas may represent former pasture (e.g. Bintliff 1992, 126).

For stone, analysis is more difficult because the material is more durable, subject to more recycling and, except for specific types, difficult to date. The repetitive and cumulative build-up of residues is usually a prerequisite for the detection of significant signals, which in turn requires a coarse-grained response to the environment, i.e. one in which activities were concentrated in particular areas (Gamble 1986; Schofield 1991, 162). Stone-reduction and manufacturing sequences in relation to site type, whether extractive or domestic, and lithic technology, whether curated or expedient, need to be taken into account (Fig. 9.6) (Gould 1980, 126). Several types of activity area can be identified: (a) A high proportion of waste debris, relatively large flakes and cores, and geologically local stone probably represent areas where stone tools were being manufactured. (b) Less waste material, a higher proportion of tools, relatively small cores and waste, and geologically foreign stone probably represent habitations where stone was brought in from a distance. (c) A lower diversity of tool types than on the other two areas and a lower proportion of waste to tools probably imply an extractive site, either for stone or food.

METHODS OF EXAMINING ARCHAEOLOGICAL DISTRIBUTIONS

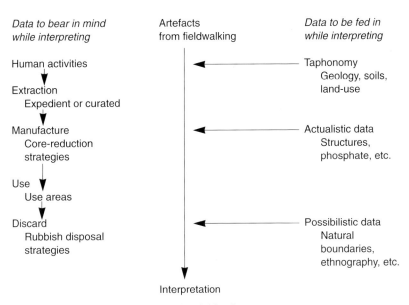

Figure 9.6 Interpretation of stone artefacts found in fieldwalking.

PHOSPHATE AND MAGNETIC SUSCEPTIBILITY MEASUREMENT

Some environments and activities are represented solely by geophysical or chemical traces. The disposal of organic materials and burning are two such activities, detectable as phosphate and magnetic susceptibility enhancement respectively. Phosphate is derived from organic materials such as bone, urine and dung. It survives in soil for millennia, strongly bound to clay; modern phosphate fertilisers do not significantly disrupt the ancient levels, being quickly dispersed. Over several hectares, phosphate survey can identify settlement locations (its original use in archaeology), although it cannot compete with aerial photography or fieldwalking over areas in excess of 10 ha (Craddock *et al*. 1985). At the intra-site scale, it can be used to detect activity areas (e.g. Conway 1983), such as rubbish tips, byres, droveways and parts of fields used by animals for defaecating. Magnetic susceptibility is based on the fact that the magnetic characteristics of a soil A horizon, in particular, can be modified and enhanced by human activities, especially burning (Thompson and Oldfield 1986). Features of both techniques are the longevity of the signals in the soil and the fact that they can reflect buried features, for example under flood loam or ploughland. So they can extend the evidence of aerial photography into areas where flood loam masks features and erosion has removed earthwork traces. Interpretation and dating need to take into account use/discard and post-deposition processes, especially soil creep, and can be done only in the context of structures and/or artefact scatters (Clark 1990).

GEOPHYSICS

Geophysical prospection is used to detect buried features, such as ditches and wall foundations, that are totally invisible on the surface except as cropmarks and can usefully

ENVIRONMENTAL ARCHAEOLOGY

Figure 9.7 Use of the gradiometer, here in the survey of a prehistoric site in Shetland.

be applied to the extension of data from above-ground survey into areas where there are no visible earthworks (Fig. 9.7). Application has conventionally been at the intra-site scale because of constraints of time, but computer technology has speeded up the process and, as with EDM survey, we can get field printouts as the survey proceeds. Even so, in the sequence of use, geophysical methods should only be applied after more extensive surveys such as fieldwalking or detailed aerial survey, or where documentary evidence suggests an early settlement (Spoerry 1992).

BIOGEOGRAPHY

The present-day distribution of plants and animals holds information about past environments, land-use and age (pp. 46–7) (e.g. Rackham 1986; 1990), a theme which should now be thoroughly familiar. In effect this principle is applicable to biological materials in archaeological contexts too. It came to be appreciated by archaeologists studying the historical periods in the UK in the 1960s with the technique of hedgerow dating, based on the principle that older hedges have more species of shrub and tree than younger ones (Pollard *et al.* 1974). A variant is the identification of hedges which have been carved out of woodland as shown by certain herbs characteristic of old woodland, like dog's mercury (*Mercurialis perennis*). Tying in the evidence with field archaeology, charters and maps is especially rewarding, although there are many problems such as the planting of multiple-species hedgerows in some English counties and the flimsiness of some parish boundaries known to be ancient (Aston 1985). Woodland plants alone – bracken (*Pteridium aquilinum*), bluebell (*Scilla non-scripta*) and dog's mercury – in otherwise open land are considered as evidence of past woodland, although bracken also spreads into abandoned arable.

In England, Rackham (1990) has studied the age and management systems of woodland from the species of tree, surviving physical evidence of the pattern of felling and cropping, their age, and the nature of the ground vegetation. Dendrochronology identifies periods of harsh weather and intensive woodland use such as the intensive cutting of leafy branches for browse. Maps, then documents and, before those, charters, placenames and ultimately field archaeology take the story back into prehistory. This is where Rackham says British woodlands often began and that Anglo-Saxons and other Migration Period settlers made little impact, and the same may apply to other kinds of un-intensively used land like meadows and commons. In Greece, the distribution of woodland and open land today is a reflection of the varying intensity of land-use in Classical times (Zangger 1992): there is dense woodland in mountainous border areas between Boeotia and Attica which were unused, while in the territories of these states, which were intensively used for agriculture, the plains and mountains alike are bare. In the Faeroes, the distribution of certain earthworm species is correlated with areas of old cultivated land (Enckell and Rundgren 1988). Particular species of animal and plant indicate ancient woodland or established grassland, and unusual abundances of certain species in environments to which they are ill-adapted suggest disequilibrium. Relic animal and plant populations, especially those of Boreal-Alpine distribution, suggest the former widespread occurrence of particular climates, and global distributions of major animal and plant groups can inform us about continental movements in the very distant past. Hedgerow dating has a deep and distant pedigree.

Conclusions

These methods are an integral part of all archaeological research. They are not subordinate to more conventional time-depth studies. Data from nucleated sites – dense concentrations of remains such as settlements, monuments and caves – preserve only a partial record, and must be complemented by those from areas of dispersed remains. Use begins at the largest spatial areas of land and works its way down to the smallest. But there is no strict hierarchy of methods with each in a set place for a set function; phosphate analysis can be used early on as a technique to locate settlements and later to identify activities within buildings. Especially important is that distributions should be traced into blank areas as into woodland or under alluvium. The deeply stratified and richly organic Upper Palaeolithic sites at Stellmoor in the Ahrensburg valley, Germany (Box 11.1), were located in just this way, with *extensive* surface survey of the valley sides on the Stellmoor-Hügel, followed by *intensive* excavation in the valley bottom (Rust 1943).

TEN

ANALYSIS AND USE OF SOILS AND SEDIMENTS

In chapter 3 we discussed some of the processes in the formation of soils and sediments and some of the ways in which they could be grouped. Now we turn to their field and laboratory study, and their archaeological significance.

MAPPING

Soil maps, at various scales and degrees of detail, are published for much of Europe and North America, and these are an invaluable starting point. For example, in the UK, soil maps at a scale of 1:63,360 or 1:25,000 are available for most of England and Wales, with 1:63,360 and 1:50,000 maps for much of Scotland (Avery 1990, 43–60). Drift geology maps at the same scales give data on sediments, many of which are quite recent and directly relevant to archaeology. In the absence of published maps, or if greater detail is needed, it may be necessary to survey the modern soil cover of the study area.

Recognising and distinguishing soils and sediments is the first step (Box 10.1). Soils are recorded from soil pits, cores and the surface where there is erosion or ploughing, to provide a general map of the distribution of soil types which can be used to predict those in unsampled areas (Shackley 1975, 23; Davidson 1980). After this, we can take short cuts for other parts of the area where a relationship between soil type and topography or vegetation has been established; aerial and satellite photography are valuable here. Localised areas of particular soil types and boundaries between soil units can be mapped in more detail, but this approach cannot be used instead of a grid or random survey because broad patterns will be missed. For sediments, since they are often localised, a judgemental strategy is used, mapping in relation to topography, vegetation and along catenas: for example, transverse and longitudinal sections of valleys (Fig. 9.1). Sediment basins which are invisible as surface topography can be located from air photographs, often showing as vegetation or soil-marks of buried river channels, infilled periglacial features, or colluvium.

Maps of different characteristics are thus obtained or compiled. Soil characteristics may be refined by subsequent laboratory analyses, but important attributes are thickness, distinctive textures, gleying (Box 10.2), iron pans, multiple profiles (Fig. 3.7), and variation in these along catenas, especially of slope (Fig. 4.5). Soil and sediment units are ultimately named but this cannot be done unless they are mapped as tabular entities and in relation to topography (Bowen 1978).

ANALYSIS AND USE OF SOILS AND SEDIMENTS

Box 10.1 Distinctions between soils and sediments

Ploughwash = Colluvium

Periglacial colluvium

The Pitstone Soil

Involutions

Section through a dry valley (cf. Fig. 12.9) in chalk showing a sequence of colluvial deposits separated by a Late-glacial soil (Rose et al. 1985); scale in 10 cm intervals.

In the field, the distinction between buried soils and sediments may be clear, as with buried soils beneath archaeological monuments (Fig. 12.2) or sediments in a lake basin. But in many cases the distinction is less so. Even with obvious soils, there is always a component of sediment derived from the material from which the soil formed and sometimes from wind-blown or colluvial material, but so slight as not to appear as a discrete layer.

A buried soil in field section shows as a thin, distinctive and continuous layer, usually darker (more organic) than the materials above and below. Buried soils have a crumb structure, have weathering horizons of different chemical and textural composition, but do not show stratigraphy. There may be a clear surface and a downward gradation of increasingly lighter colour or other properties such as mottling. Clay layers pose problems since they may be of a variety of origins – as a sediment in the subsoil, as a sediment washed into the soil during its formation, as a soil weathering horizon, or as a soil illuvial horizon formed by clay translocation (lessivation). Micromorphology can separate these different origins as we shall see in the case of the Pech-de-l'Azé (p. 124).

Sediments are often thick and of variable thickness, show stratigraphy, are often paler and less organic than soils, and do not have a crumb structure. They do not show weathering horizons, although zones of iron staining may cut across the layering. These distinctions are not always clear, especially in Pleistocene deposits where a sediment may be derived from a pre-existing soil by solifluction. Layers of clastic material within purely organic soils or soils formed from biogenic deposits like travertine are a clear indicator of a sedimentary interval.

Biological inclusions are an important clue. Soils often show a gradation in abundance of biological inclusions from high at the surface to low in the subsoil, and there is often a high peak of abundance at the surface itself (Fig. 7.8). The assemblages often comprise coherent ecological groups. Assemblages from clastic sediments, in contrast, especially where water-lain, are often ecologically mixed, with land and aquatic species in the same unit (pp. 40–1, Fig. 3.11).

FIELD DESCRIPTION

Field description can only be learnt in the field, although there are several useful guides (Shackley 1975, chapter 2; Courty *et al.* 1989, chapter 3; Limbrey 1975; Trudgill 1989). For soils, description is best done from a pit (Figs 10.1 and 10.2), for sediments from sections (Box 10.1) or cores; different kinds of corer are described by West (1972, 99–108). Many properties, especially texture, are best described relative to adjacent horizons in the first instance because whether a horizon is more or less clayey, for example, than one below or above is often more important in terms of processes than are actual proportions of clay.

Horizons and layers are characterised first by texture, for example clay, silt or loam (Table 3.3), and then by various qualifiers like structure and consistency. Stones are recorded in terms of abundance, size, shape, angularity and lithology. For example, the presence or absence of foreign stones (erratics) and the alignment of clasts along the direction of transport are important criteria in separating solifluxion deposits from glacial till. Colour is recorded with reference to a standard soil colour chart. For example, the widely used Munsell system (1994) gives a formula and named colour (Fig. 10.1). The level of weathering – oxidation and hydrolysis – is indicated by 'hue'; this lies towards the brown (10YR) end of the scale in oxidised soils of temperate climates, the red (7.5YR to 10R) end of the scale in the more oxidised and dehydrated soils of Mediterranean and other semi-arid climates, and towards the yellow and olive (Y) end for reduced, gleyed soils. Orange/grey mottling is an important indication of semi-waterlogging (Box 10.1). 'Value' indicates the degree of blackness, usually an index of organic matter, although it can be affected by some clays, variations in moisture content, and minerals like pyrolusite

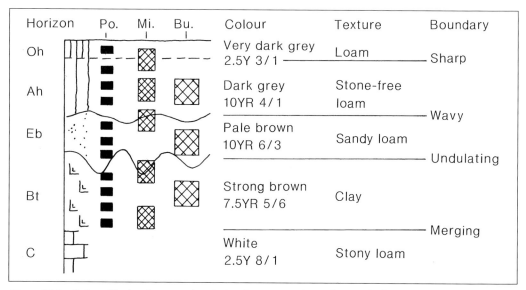

Figure 10.1 Soil profile showing some descriptive characters and sampling methods; Bu. = bulk samples, Mi. = micromorphology samples, and Po. = pollen samples (cf. Figs 7.6 and 10.2; Courty et al. 1989, fig. 3.9).

(manganese dioxide). 'Chroma' is a measure of the intensity of weathering. For example, the Oh horizon in Fig. 10.1 has a hue of 2.5Y, a value of 3 and a chroma of 1, written as 2.5Y 3/1, very dark grey. Other features can often be inferred from simple tests or observations, such as humus (dark, strong crumb structure), calcium carbonate (fizzes with dilute acid, presence of bone and shell) and iron oxides (orange and/or grey). Anthropic materials like mortar, plaster, clay, charcoal, pottery and tile are recorded at the same time. In sediments, bedding, as reflected by vertical and lateral changes in texture, sorting and clast composition (Table 3.3), and in soils, horizons and their boundaries are recorded (Fig. 10.1), along with distinctive structures of periglacial, biotic and other origins, e.g. cultivation marks, rootholes and cracks (e.g. Table 7.1).

Bulk Laboratory Analysis

Samples for bulk analysis must encompass vertical and lateral variation. Lateral sampling is usually closer for soils, especially surfaces, than it is for sediments because of the likely greater degree of variation in the former (Fig. 7.6). It is important that samples are confined to clear layers or horizons and do not cut across them (Fig. 10.1). Bulk-sample analyses are useful for measuring chemical properties like organic matter, C:N ratio (p. 31), calcium carbonate, pH, iron oxidation state, phosphate and heavy minerals (Courty *et al.* 1989; Limbrey 1975), but they are specialised and often expensive, and should only be done with specific questions in mind. In buried soils, organic matter, pH and C:N ratio are likely to have been in a state of flux since burial, and may not be especially meaningful. A severe problem of bulk analyses is that they do not identify structure. The difference between clay occurring as linings of pores or as fine laminae is not revealed. Nor are particles which were deposited as aggregations, for particle-size analysis records only individual grain distributions; for example, a sediment made up of silt- or sand-size aggregates of clay particles laid down in a moderate energy depositional environment will appear to be a clay, with implications of low energy deposition (see pp. 39–40 and Table 3.2 for different energies of deposition).

But some analyses are useful and can only be identified from bulk samples. Grain surface texture under the scanning electron microscope can indicate the depositional environment, especially for sands (Tankard and Schweitzer 1976); grains are etched differently according to conditions, especially the distinction between aeolian, water-worn and chemically weathered grains. Mineral magnetics and other magnetic properties are a useful indicator of inwashing of old topsoil and of burning; they can distinguish between the inwash of topsoil as might result from human activity and that of deeper sediments brought about by river channel scour, and can separate different types of human activity such as light and deep ploughing, quarrying (Thompson and Oldfield 1986; Lageras and Sandgren 1994), and natural and artificial fires (Bellomo 1993). Heavy metal analysis can identify episodes of inwash into alluvium as a result of mining activity (Lewin *et al.* 1977; Macklin 1985). Loss on ignition indicates the relative proportions of mineral and organic materials and is useful as an index of the amount of mineral inwash into peaty deposits and soils (e.g. Mills *et al.* 1994), especially in conjunction with mineral magnetics.

ENVIRONMENTAL ARCHAEOLOGY

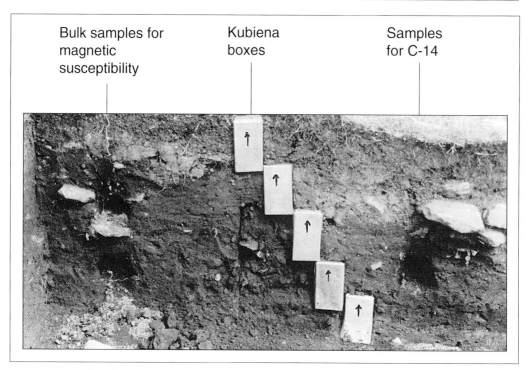

Figure 10.2 Soil profile in Shetland showing categories of sampling; arrows on Kubiena boxes indicate orientation.

Micromorphology

Micromorphology (Courty *et al.* 1989; Macphail and Goldberg 1995) is the study of undisturbed material in thin section and it allows features of soil horizons, sediment structures and boundaries to be examined under the microscope. These features are perhaps closer to the processes which formed them than are other kinds of data. Samples for micromorphology are taken contiguously, and especially *across* boundaries (Figs 10.1 and 10.2). Blocks of soil are taken in Kubiena boxes from the profile, noting the vertical orientation. The soil is impregnated and consolidated with resin, a smooth surface cut and ground, and this affixed to a glass microscope slide; the other side is then ground down until 25–30µm thick, sealed with a coverslip and examined under a polarising microscope.

Micromorphology enables soil and sediment structures and the arrangement of particles, pores and linings to be studied in their position of life. Modern analogue studies from experiment and ethnoarchaeology have shown particularly strong and clear links between specific processes and their micromorphological expression (p. 125) (Macphail and Goldberg 1995). Particle sizes and aggregation, grain shape, mineralogy and biological features can be quantified, and remains, burrows and droppings of the soil fauna can be recorded. Limbrey (1992) discusses the biological origins of calcigenic floodloams, possible agencies being blue-green algae, lumbricid earthworms, calcification of decaying plant roots, and concentration in mite tissues; only micromorphology can

ANALYSIS AND USE OF SOILS AND SEDIMENTS

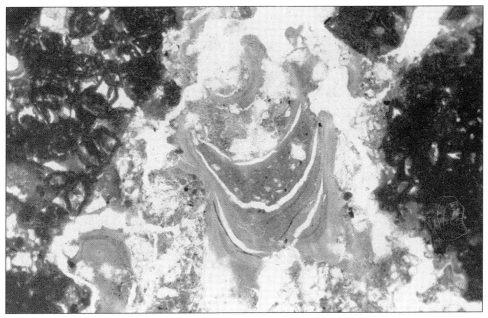

Figure 10.3   Photomicrograph of a soil thin-section from a buried soil beneath a Bronze Age barrow at Deeping St Nicholas, Lincolnshire; from the lower horizon of an argillic brownearth showing successive crescentic infilling of a void with oriented clay and silt. Frame width = 4 mm. (Photo, C. French.)

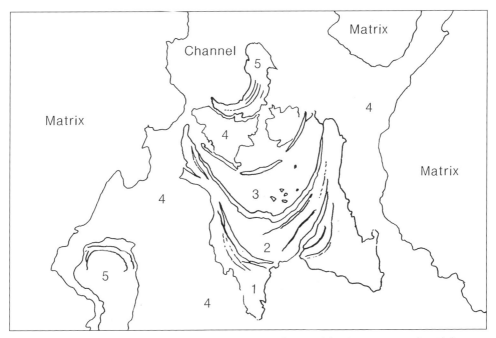

Figure 10.4   Annotation of photomicrograph in Fig. 10.3. 1 = laminated, birefringent, pure or limpid clay; 2 = laminated, birefringent, dusty/impure or silty clay; 3 = non-laminated, dusty clay with inclusions of quartz; 4 = non-laminated, dusty/dirty clay or silty clay with amorphous organic matter, partially reworked by soil fauna; 5 = alternate laminae of pure and dusty or impure clay; matrix is iron-impregnated. For sequence, see pp. 124–5.

123

Table 10.1. *Pedozones in the infilling of a hypothetical tree-throw pit beneath an archaeological earthwork; 1 is the earliest (cf. Fig. 10.5)*

Pedozone	Material	Soil profile interpretation
1	C horizon	Original, pre-tree-hole, profile
	Bs horizon	
	Ah horizon	
2	Illuviated clay, Bt material coating peds in tree-hole material	Clay illuviation caused by tree-throw or cultivation afterwards
3	Root channels with dusty coatings	Further disturbance, possibly associated with cultivation and burning
4	Charcoal fragments with illuvial clay coatings	Inclusions of human occupation debris and further illuviation
5	Calcitic silt as outermost coating of peds	Post-burial downwashing of water and solutes into the buried soil, possibly caused by ploughing and exposure of damp topsoil, clogging up soil pores, and all leading to soil impoverishment

identify these processes of biomineralisation. Thin sections of tufa often show filaments of blue-green algae which were responsible for its deposition (Evans and Smith 1983, pl. 11). Micromorphology also records anthropic inclusions like charcoal, straw, phytoliths, pottery and daub, and it has been used in the characterisation of fire horizons in Middle Palaeolithic deposits (Rigaud *et al.* 1995). The characterisation of the various states of clay is important, especially in distinguishing soil horizons from sediments. Goldberg (1992, 159) identified red-clay units in an Upper Palaeolithic cave at Pech-de-l'Azé II in the French Perigord as solifucted Tertiary deposits formed in cold conditions rather than soils with illuvial clay formed in warm conditions. The differences are subtle. Clay coatings (cutans) formed in illuviation are fine, uniform and micro-laminated. The ones at Pech-de-l'Azé were thicker, poorly sorted and poorly oriented, and they coated grains and aggregates rather than lining voids which is more usual with cutans. One particular kind of coating is known as a *dusty clay coating* because it is made up of clay, silt and soot or charcoal particles. These coatings, often micro-laminated, indicate soil disturbance which can, for example, derive from vegetational clearance and cultivation, breaking up of soil peds at the surface by rain, and downwashing of the material into lower horizons; they are thus good indicators of human activity. They also damage the soil, clogging voids, impeding drainage and can lead to waterlogging and erosion or peat growth (pp. 34–5).

Micromorphology allows time sequences to be identified from a single thin section (Macphail 1991; Goldberg 1992, 146; Limbrey 1992). The one illustrated here (Figs 10.3 and 10.4) suggests disturbance (2–3 and 5) of an argillic brownearth or woodland soil (1), probably prior to and during the construction of the barrow which ultimately buried it; the soil was then subject to reworking and bioturbation (4); subsequently, alternate wetting and drying led to iron impregnation of the matrix, probably brought on by the post-barrow growth of peat, rising groundwater and flooding, followed by more recent drying.

More complex sequences have been described from tree-throw pits, a hypothetical example being illustrated here (Fig. 10.5 and Table 10.1). The different stages are called *pedozones* (Macphail 1991; 1992). In pedozone 1, a mixture of discrete fragments of different horizons of a soil profile occur as a micro-scree. Although the original soil may have been destroyed by truncation and erosion, its profile can be reconstituted from these fragments. In pedozone 2, all these fragments have been coated with clay materials from later lessivation, indicative of woodland clearance. In pedozone 3, root channels penetrating the matrix are lined with dusty clay coatings, indicative of cultivation, while fine charcoal in the root channels and its dispersal indicate burning and biological activity. In pedozone 4, illuvial clay around the charcoal suggests further illuviation (cf. Goldberg 1992). This was followed by translocation of calcitic silt through the burial matrix into the buried soil (cf. Limbrey 1992), pedozone 5.

In both these examples it was possible to separate processes of pre-burial pedogenesis from those occurring at the time of disturbance and burial, and both of these from processes of post-burial diagenetic origin, as well as defining separate stages within these. Micromorphology is becoming an essential tool in soil studies, but each case must be interpreted in relation to modern analogues (chapter 14), against a background of the macromorphology and soil/sediment distribution, and the distribution of sites and artefacts.

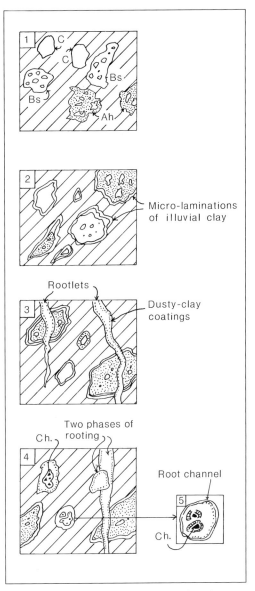

Figure 10.5 Sequence of pedozones as based on a single thin-section from a tree hollow; for interpretation, see Table 10.1. Fragments of soil material are: Ah = Ah horizon; Bs = Bs horizon; C = C horizon; Ch. = charcoal; oblique hatching = matrix. The five frames are a sequence of development reconstructed from a single thin section.

Box 10.2 The prediction of site distributions

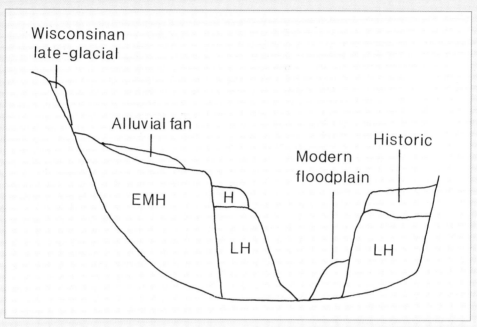

A succession of terraces from the central Des Moines valley. For key, see Table.

In a study in the Upper Midwest (Bettis 1992), river terraces were characterised as 'age-morphologic groups' (AMGs) on the basis of radiocarbon age and altimetric and stratigraphic relationships. The AMGs were then characterised sedimentologically and related to the regional lithostratigraphic units.

Criteria by which the AMGs could be recognised in the field without excavation were then established. The use of altimetric data was one possibility, but this is not foolproof because younger terrace material overlies older terraces in some areas. A useful property of river terrace soils is their increasingly stronger profile development on successively higher (=older) terraces. This has been

AMG	Age BP	Geomorphology	Surface soils
Wisconsinan Late-glacial	>10,500	Benches and terraces	–
EMH–early to mid-Holocene	10,500–4000	High terraces, alluvial fans, colluvial slopes	Mollisols or alfisols – A-E-Bt or A-Bt
LH–late Holocene	3500–450	Low terrace complex and floodplain	Mollisols or inceptisols – A-Bw
H–Historic	<450	Floodplain	Entisols – A-C (no B)

The prediction of site distributions

shown, for example, with Holocene terrace soils in Wisconsin (e.g. McDowell 1983; Ferring 1992, 8) and central Europe (Schirmer 1988), and it is the case in the present study. There is a positive correlation of properties like oxidation state (assessed from field characteristics of colour and mottling) and soil structure with the age of the terrace surfaces. On this basis, four 'landform-sediment assemblages' (LSAs) were established. These roughly correlate altimetrically with particular terraces, but the soil properties are the key to field recognition.

Because of their different ages, the LSAs are likely to have different archaeological distributions, which can then be predicted as shown in the following table of preservation potential for buried cultural deposits. The oldest distributions are fragmented, occurring in the remnants of the oldest terraces; the latest occur only in floodplain deposits, although with a wider surface distribution. The LSA data also allow estimates of volume distributions to be made, so that these can be taken into account in sampling programmes. Additionally, they permit an assessment of the amounts and locations of land of each stage that have been destroyed, thus enabling archaeological distributions to be understood better in these terms.

Cultural period	EMH Alluvial fans	LH High terrace	Historic Low terrace	Floodplain
Paleoindian	+ +	+ (late)	–	–
Early and Middle Archaic	+ +	+ +	–	–
Late Archaic	+ +	+	+ +	–
Woodland	+ –	–	+ +	–
Oneota and Great Oasis	+ –	–	+ +	–
Historic	–	–	+ –	+ +

– = not possible, + – = low potential, + = moderate potential, + + = high potential

SOILS AND SEDIMENTS IN ARCHAEOLOGY

First of all, time and space units of soils and sediments are established from the field and laboratory data. Maps show the distribution of particular categories which are selected according to the aims of the research (p. 102). Type sections are established for sediment units and fossil soils (pp. 120–1), such as the Pitstone Soil (Box 10.1), which should be independently dated, for example by radiocarbon.

Taphonomy

Soils and sediments enable us to understand how the environmental-archaeological record formed, specifically with regard to site-formation histories and archaeological distributions. An immediate point is that thickness is not a direct reflection of time. Soils are relatively thinner and have a conflated record, whilst sediments are relatively thicker with a spaced-out record, and do not always form at a constant rate. There may be periods of erosion when substantial periods of time are absent (Box 10.1). Then there are the different ways in which biological materials are incorporated into soils and sediments, which entail different sampling and interpretational procedures, as with the distinction between soil and peat pollen analysis (Box 6.1, p. 135). Differences in lateral variation are also important. Soils and sediments can reflect climate but this is often strongly modified by the context and in the Holocene by human activities, so the different degrees of lateral variation between soils and clastic sediments entail different sampling procedures (Fig. 7.6). More widely, the patchy spatial distributions of fossil soils and sediments, their preservation in some areas, their absence in others, and the varying surface distributions are relevant to the distribution and interpretation of field archaeology (Figs 7.9 and 7.10). Variability is particularly relevant to river valleys where complex soil and sediment distributions are formed by the shifting and downcutting of the river (pp. 160–1). Archaeology can be related to specific terraces, and this is useful not only in explaining distributions but also in predicting them (Box 10.2).

Soils and past land-use

When our study area has a long history (and prehistory) of cultivation, it is wrong to assume that the present distribution of soil types reflects that when the area was first settled. The more humic upper horizons (topsoil = epipedon), especially, may have been more humic and uniform than today. As cultivation proceeded, increased incorporation of subsoil into the upper layers could have produced quite a different distribution of soil types, for example by accelerating the loss of nutrients from soils on acid sands or inclining upland soils to waterlogging (Fig. 3.5). Alternatively, manuring and other interventions may have greatly increased the organic content and potential fertility of some soils (Davidson and Simpson 1984; Groenman-van Waateringe and Robinson 1988). This is why a part of archaeological fieldwork is the study of relationships between soil types and archaeological distributions and an interpretation of their causes.

Archaeologists use a range of soil properties, from detailed distinctions, e.g., 'sandy' and 'loamy-sandy' soils (Bakker and Groenman-van Waateringe 1988), through numerical (Davidson 1980) and descriptive land classification grades like 'very poor to very good soils' (Gaffney *et al.* 1995; Gardiner 1980), to general terms like 'modern arable', 'productive land' or 'exhausted soil'. They are usually based on a supposed relationship between subsistence farming and its dependence on particular types of soil. Positive or negative correlations can be assessed visually or statistically. Relationships are particularly persuasive when we can show a linear relationship between settlement size and area of such land (e.g. Ellison 1980) or where there are close relationships between concentrations of settlements and good agricultural land (e.g. Fig. 15.5).

However, the causes of relationships between archaeology and soil type are difficult to demonstrate, even where post-depositional factors can be discounted. Correlations can be with climate or vegetation which themselves are correlated with soil type, while productive soils often coincide with areas which are suitable for agriculture because of shelter, access to water, communications and harbours. Even when we have the evidence of buried soils, identifying their past productivity is difficult because of loss of organic matter and nitrogen through biological activity after burial, and changes in pH and nutrient levels. Crumb structure as viewed in thin section is probably the best guide. This is especially a problem in temperate regions but may be less so in semi-arid areas, where there is less post-depositional biological activity.

Soil and sediment histories

Each soil and sediment unit refers to an environmental stage in terms of soil formation and sedimentation respectively. We can date the boundaries of a sequence, for example with radiocarbon, and characterise its stages in more detail using biological data, but the basic scheme is pedolithostratigraphic and must be interpreted in terms of soil and sediment processes in the first instance. In a temperate-climate lake (Fig. 10.6), organic muds form as a local response to dying vegetation and as a regional response to the eutrophic conditions engendered by the climate. A layer of minerogenic clay then reflects runoff and erosion from the land surface in the catchment, possibly as a result of farming in the catchment. It could be tied in by pollen analysis of the sediments themselves, but wider correlation with other archaeological data like settlements must be done by radiocarbon dating. In our example the breaks in the litho-, pollen and archaeological stratigraphies do not correlate entirely. There is a time-lag between farmers moving into the area and widespread modification of vegetation. There is also a time-lag between cultivation and the inwashed clay, which even then seems to equate more with abandonment. This non-matching of different stratigraphies is normal: we must have independent dating of each before correlations can be made.

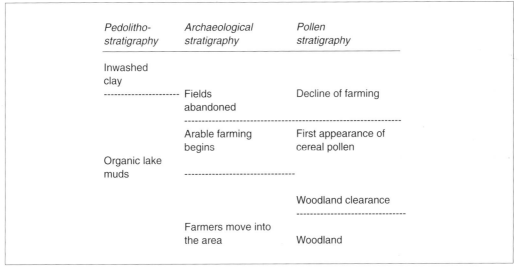

Figure 10.6 *The non-equivalence of different stratigraphies from evidence around a lake and in its sediments.*

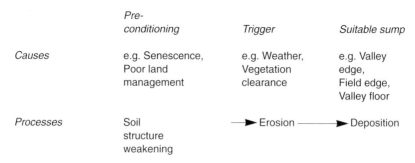

Figure 10.7 Different levels of causes and requirements of soil erosion and deposition.

Causation, however, is still a problem (just as we saw with spatial correlations), especially as there is a chain of events from distal origins in the ecosystem as a whole like climate change, through an effect on soil processes like a fall in pH, to a proximal trigger like agriculture which then brings about erosion (Fig. 10.7).

Much evidence is circumstantial, in that it is based on correlations between sequences, and the debate is often polarised between climate and humans as the driving forces (Bell and Boardman 1992; Needham and Macklin 1992). Causation can only be demonstrated if we can establish mechanisms, especially those which might render a soil more susceptible to erosion.

For example, overbank alluvium is formed in a number of stages (Fig. 10.8):

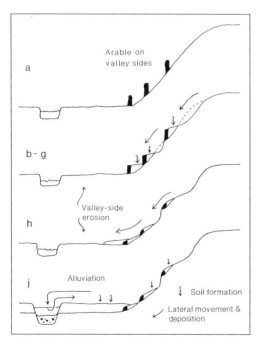

Figure 10.8 Stages in the formation of fine overbank alluvium.

(a) Increased susceptibility by weakening of the soil crumb structure. This is the primary prerequisite for erosion; soil with a strong crumb structure is very resistant to erosion even if the surface is bare and the slope steep. Crumb-structure weakening can be caused by human land-use, climate change, and changes intrinsic to the soil and vegetation.

(b) Erosion. This can be a consequence of crumb-structure weakening and a vegetation-free surface. It can be helped by particles, liberated into the soil from the weakened crumb structure and sometimes including charcoal, clogging the spaces in the remaining crumbs and thus impeding drainage and accelerating runoff.

(c) In other cases, erosion is triggered by external changes like an increase in rainfall or a change in ploughing season. But these, unless they are also responsible for weakening the soil crumb structure in the first place, are not the *primary* cause.

(d) Alteration of the size, shape and surface of particles during erosion.

(e) Deposition of eroded material in a land catchment. This can be against field edges (Fig. 10.8) or in a dry valley (Fig. 12.9), when it takes on depositional structures like clast orientation and bedding.

(f) Alternatively, complete removal of eroded material from the valley system. There is not always harmony between erosional histories and valley sediment sequences.

(g) Post-depositional weathering (pedogenesis) of deposited material. A new crumb structure is superimposed on the depositional features and any of the original soil crumb structure that remains.

(h) Land abandonment leads to the dilapidation of terracing and a new sequence of erosion and deposition, now onto the edge of the valley bottom.

(j) As a result of relaxed management, the river then cuts into field edges and washes the colluvium downstream, redepositing it as overbank alluvium. New depositional structures are formed and new pedogenesis ensues.

How do we work our way back from alluvium to our original soil on a hillslope in order to isolate the various causes of weathering, erosion and deposition? This is the area of interpretation, to which we return in Part IV, but a few comments are useful here. First there is the circumstantial evidence of correlations between alluvial and other sequences like episodes of river channel movement (Needham and Macklin 1992) or human settlement histories (e.g. Bintliff 1992). Second, there is the topography and the relationship of the sediment sump to archaeology and natural features. Is it likely the hillslopes were cultivated? Are there traces of terracing? Third, there is the content of the sediments: a strong component of anthropic materials indicates the presence of humans, while biological materials can indicate the environment of deposition, including earlier terrestrial ones (chapter 11). Finally there are the sediments themselves, especially their micromorphology, and we have already noted that different types of weathering and deposition give rise to particular fabrics, and that often a sequence of these can be preserved in a single sample.

This discussion has shown the importance of using a variety of scales. The chronological pattern of alluviation in a region, and its relation to archaeological events such as demographic change, is as much evidence as a single soil micromorphology thin section. We return to this point in chapter 15.

Eleven

BIOLOGICAL INDICATORS

In this chapter, we examine the groups of organisms which are used as environmental indicators. Some general points are made first, followed by examination of organisms by taxonomic group. This structure arises in part from the tendency for different projects to be based on particular taxonomic groups rather than on a broader reconstruction of biotic communities, and in part from the need to subdivide a large topic.

GENERAL ISSUES

In using plants and animals as indicators of past conditions, we are using direct evidence of past biota, but indirect evidence of other aspects like climate (pp. 12–13). To some extent, this is an exercise in identifying the fossils and interpreting them in the light of modern community ecology. However, several issues complicate this process. A fundamental point is the application of the species concept to ancient material (p. 15). A specimen recovered from sediments after thousands of years may no longer bear enough features to permit its identification to an extant species. This may be because a shell, for example, is incomplete or distorted (Fig. 11.1) or because a particular group of species is separated by features not preserved subfossil, or because the specimen is of a species now extinct.

Identification may therefore be restricted to a higher taxonomic level. The question also arises as to how far ancient material should be equated with extant species at all, given that the holotype of some species will include morphological features which cannot be seen on ancient specimens. This is seen in the identification of pollen grains, where specimens of one specialised plant structure are named according to a taxonomy founded on descriptions of whole plants. Pragmatically, we have to call the fossils something, and often it is possible to be confident that the specimens are bits of a

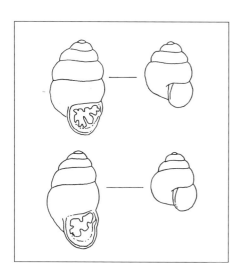

Figure 11.1 Shells of Vertigo antivertigo *(above) and* V. moulinsiana, *showing clear differences between the adults (left) but close similarities between the juveniles.*

particular species. However, the use of 'form-fossils' is sometimes more honest. Coles (1990) has used this approach in naming spores and other structures in pollen preparations. The taxonomy is 'user-defined' and expedient. Similarly, Hillman *et al.* (1996) have proposed criteria and taxonomic recommendations for remains of wheat (*Triticum* spp.).

One also has to decide what taxa are likely to be represented in an assemblage. In Newnham's (1992) work at Otakairangi, Northland, New Zealand, pollen attributable to the tree taxon *Nothofagus fusca*-type (which includes *N. fusca*, *N. truncata*, and two sub-species of *N. solandri*) was attributed to *N. truncata* in part on the basis of the morphology of the grains and in part on biogeographical considerations. An implicit decision is thus taken that certain species are unlikely to have been present at a given place and time. This important compromise is given little consideration, though Driver (1992) raised the issue with respect to vertebrate remains. He observes that a large bovid femur fragment from a 3,000-year-old site from the Canadian plains will almost certainly be identified as *Bison*, whereas the same specimen from an historic period site in the same area, where domestic cattle could also be present, would be recorded as *Bos/Bison*.

In order to justify any assumptions about past distributions, we need to understand the possible effects of human activity and climate. For example, the present distribution of the yellow-necked mouse, *Apodemus flavicollis*, in Britain is markedly disjunct, and the species shows a strong association with deciduous woodland (Corbet and Southern 1977, 217). It is thus probably correct to argue that woodland clearance has fragmented a formerly more widespread range, and so the recovery of ancient specimens of *A. flavicollis* from outside its present range comes as no surprise (Yalden 1983). Other instances are less simple to interpret. If specimens of a thermophilous woodland bug are found further north in Europe in the early Holocene than in later millennia, does this indicate a fall in mean temperatures, is it a consequence of woodland clearance, or did the bug just become uncompetitive at the margins of its range as the woodland invertebrate community developed (pp. 18–19)? Competitive exclusion is often invoked in neoecology (Begon *et al.* 1990, 251), yet it is rarely considered as a factor for past communities. Similar questions arise with the 'elm decline' recognised in pollen diagrams across Europe, explanations for which include climate change and human activity (Bell and Walker 1992, 160). Disease has also been mooted (Peglar 1993), but little consideration has been given to the possibility that the elm decline might indicate the Holocene deciduous woodland succession passing from one stage to another (see Smith 1965 for an early exception). The discussion has centred on factors external to the community.

Human activities create another headache, especially in regard to synanthropic behaviour. Some organisms are encouraged by the microhabitats created by human settlement, or by the reduction of competing species or predators; indeed, one of us has argued that this process has facilitated domestication of some animals (O'Connor 1997; and see also Zeuner 1963). The rabbit, *Oryctolagus cuniculus*, has multiplied almost exponentially on the sheep ranges of Australia, an exotic wild animal benefitting from the habitat maintained for an exotic domestic animal. Species which are obviously synanthropic can be readily accommodated in environmental reconstructions, but it is another matter to be confident that a particular species was not synanthropic in the past in a habitat which humans do not maintain today. Can we be sure that species do not expediently adopt synanthropic habits? Humans have also caused the reduction and extinction of animals and plants, as with the clearance of Holocene woodland in large tracts of Eurasia, fragmenting the distributions of many herbaceous plant and shrub species. The effect of human populations on the past distribution and abundance of organisms and the importance of grasping how little we understand of past associations and change could be laboured. In using biological indicators of past environments, we have to keep in mind the trophic relations of humans: we are broad-spectrum omnivores, and thus have an impact at numerous points in the food-chain. We are adept at engineering habitats and naive in our attempts to control them. Human food-procurement or the extraction of raw materials can have unexpected effects and produce associations in our ancient samples which are quite unfamiliar to us.

Plant Microfossils

Pollen

The study of pollen from ancient soils and sediments dates back to the beginnings of modern archaeology and is at the heart of many collaborations between archaeologists, geographers and botanists (Berglund 1986; Dimbleby 1985; Faegri and Iversen 1989; Moore *et al.* 1991). Pollen is produced in greater or lesser amounts by all spermatophyte plants. Mature grains are often produced in prodigious quantities and are dispersed variously (Box 6.1), forming a high proportion of the airborne dust which settles on the ground, in lakes and in rivers. Pollen grains of different taxa differ in size and appearance. The gamete-bearing tissues are surrounded by a layered 'shell', the intine and exine of which may be more or less distinctively sculptured. Identification thus uses overall size and shape (spherical, ovoid), the presence or absence of major features such as air sacs, grooves and pores, and the appearance of the grain surface. Pollen grains are typically of the order of 20–100 µm in diameter. The exine is largely composed of a polymer, sporopollenin, which resembles a polymer of carotenoids and carotenoid esters with oxygen. It is resistant to decay and occurs generally in the protection of propagules (= bodies of plant propagation); traces have been claimed in some of the oldest sedimentary rocks. In burial, the cellulose entine and polysaccharide contents of the grains are destroyed, leaving the exine fossil. Corrosion and degradation may occur through localised attack by micro-organisms or direct chemical oxidation. Damage may also occur through the deposition of minerals such as iron pyrite and marcasite within the grain (Delcourt and Delcourt 1980). Preservation is best in acidic, anoxic media such as peat or close-textured limnic sediments, and poorest in open-textured, calcareous sediments and soils (Box 7.1).

We have already reviewed the movement of pollen from plant to deposition (Box 6.1). While such work has shown the poor spatial resolution of many pollen analyses, it has allowed better definition of the catchment represented by a particular pollen core

(Bradshaw 1991). The use of multiple cores to examine pollen concentrations in three dimensions within a sediment gives data on spatial variation. Extraction and counting techniques can also allow conversion of counts per microscope slide to counts per gram of sediment, usually by incorporating into the sample a known quantity of an exotic spore, *Lycopodium* sp. or *Eucalyptus* pollen, and normalising the data with respect to its concentration. Thus at Little Hawes Water, Lancashire, Taylor *et al.* (1994) were able to relate spatial variation in the concentration and taxonomic composition of pollen samples to the location and intensity of human activity within a wetland catchment.

Pollen studies are conventionally subdivided into regional studies, which seek to examine the past pollen rain of a substantial area (many hectares) over a considerable period of time (usually several millennia), and local studies, which examine smaller pollen catchments, often with finer chronological resolution. An example of a regional study is that from Lake Zeribar, Iran, whence a long pollen record showed the pattern and rate of establishment of plant communities during the early Holocene, and the effects on the vegetation of settled agriculture (van Zeist and Bottema 1977). The spatial resolution was poor as the lake catchment was many square kilometres and particular events could only be tied down to a period of one or two centuries. Even so, the study had great long-term and widescale value. A number of British palynologists have studied deposits likely to have had smaller catchments and finer temporal resolution. Dimbleby (1985) drew attention to the value of pollen in palaeosols for this reason, and used pollen sequences from soils beneath prehistoric monuments as the basis of his classic analysis of the origins of the British heathlands. Simmons *et al.* (1993) investigated soil pollen sequences in north-east Yorkshire as a means of exploring supposed human effects on vegetation in the Mesolithic, inferring local clearance of deciduous woodland, possibly to attract herd ungulates, and probably with the beneficial effect of encouraging useful woodland-margin plants such as blackberry (*Rubus fruticosus* agg.) and hazel (*Corylus avellana*).

Pollen analysis of natural or archaeological contexts is not invariably worthwhile. During the formation of an 'anthropogenic' sediment such as a pit-fill or occupation surface, allochthonous pollen may be introduced in such quantities as to swamp any record of the autochthonous vegetation or regional pollen. For example, herbs collected whilst in flower for culinary purposes can introduce huge quantities of pollen of a single species which may be quite exotic to the site. It is possible to recognise this by the presence of immature grains and clumps of grains, features of pollen still forming in the anthers. This line of argument was used at Shanidar, Iran, to justify an interpretation of pollen in the sediments around a Neanderthal skeleton as reflecting the deliberate emplacement of flowers (Leroi-Gourhan 1968). On urban sites, however, the potential sources of the pollen in a deposit are numerous, and its interpretation is accordingly problematical (e.g. see pp. 199–200).

Phytoliths

Many genera of grasses concentrate silica within their cells as grass opals or phytoliths (Piperno 1988; Rapp and Mulholland 1992; Pearsall and Piperno 1993). They range around 25–250 µm in length and are variable in shape, though rod- and dumbbell-shaped forms are common. They are inherently stable in many burial environments and can be recovered from a range of natural and archaeological sediments. Identification is difficult, as different taxa do not necessarily have distinctively shaped phytoliths, nor can particular shapes be assumed to be restricted to one genus or a few closely-related species, but some archaeologically relevant taxa, notably maize (*Zea mays*), produce sufficiently characteristic phytoliths to allow their identification to species. In European archaeology, the poor taxonomic resolution of phytoliths is inadequate for the questions of floristic or environmental change which are commonly posed. High concentrations within a sediment might, however, support an interpretation in terms of pasture habitats, or the use of straw or herbivore dung in manuring (Powers-Jones 1994). More use has been made of phytoliths in the Americas and Near East and some taxonomic advances have been made (Brown 1984; Rosen 1994). One interesting use of phytolith evidence has been in the direct recognition of diet, by analysis of phytoliths adhering to human teeth (Lalueza Fox and Perez-Perez 1994).

Diatoms

Diatoms are unicellular algae which occupy small silica chambers or frustules. They are widely distributed in the hydrosphere, with abundant freshwater and marine forms, and some which extend into wet terrestrial environments, such as moss polsters. The overall shape of the frustules differs between species, as do surface morphology and ornamentation, and they are thus amenable to a high degree of taxonomic identification. Taken with their often narrow habitat requirements, this makes diatoms a valuable source of information on specific questions concerning the origins and development of a single water body (Battarbee 1988), and much broader climatic considerations. Diatom frustules occur in most aquatic sediments, and in deep oligotrophic lakes they can be so abundant as to constitute a major part of the sediment, known as 'diatomite'. Changes in species composition through diatomites can indicate changes in water oxygenation, temperature and chemistry, and such changes may in turn reflect processes in the terrestrial catchment of the lake (e.g. Mannion 1978). Hemphill-Haley (1996) has used diatoms to enable the identification of tsunami deposits on the Pacific north-west coast of America. One interesting use of diatoms has been in provenancing pottery from prehistoric sites in Finland (Matiskainen and Alhonen 1984).

Cereal bran

A final form of plant microfossil is cereal bran. This is part of the periderm of the grass caryopsis, and consists of several layers of thick-walled cells which can be preserved in prodigious quantities in waterlogged deposits (Dickson 1987; Holden 1990). Bran is assumed to enter archaeological sediments by the alimentary pathway, that is, in faeces. Waterlogged fills of latrine pits on urban sites can yield high concentrations of brownish flecks of plant debris, the biggest fragments resembling tea-leaves. On microscopic examination, these are seen to have the characteristic cell pattern of bran, and this pattern may allow some identification of the taxa present, though seldom beyond saying that, for example, the majority is wheat bran, with a minority component of oats.

Other plant microfossils have been used as environmental indicators, notably the cysts of various groups of algae (Cronberg 1986; Coles 1990). In this brief review, we have introduced those of archaeological relevance and accept that other investigations, particularly limnological ones, may use a wider range.

Plant Macrofossils

The term plant macrofossils is usually applied to plant structures visible to the naked eye (Berglund 1986, 571–626; Dimbleby 1978; Hastorf and Popper 1988), although timber is dealt with separately here because of its use in dating and as evidence of past growing conditions. Plant macrofossils consist largely of complex polysaccharides known as *cellulose*, with or without additional reinforcement through the deposition of lignin. Cellulose and, to a lesser extent, lignin are susceptible to decay in oxygen-rich conditions, in part through direct chemical degradation, but mainly through the activities of aerobic micro-organisms. Waterlogged sediments, in which macrofossils are common, inhibit the chemical degradation and are inimical to the aerobic organisms which would otherwise reduce plant tissues to simpler, rapidly degraded compounds. Plant macrofossils occur in a range of burial contexts, from anoxic peats and lake muds, through waterlogged, and therefore anoxic, archaeological deposits like pits, wells and fine silt in the bottom of ditches in areas of high groundwater, to fully oxic contexts but in these only as charred remains (Box 7.1). Some parts are more commonly recovered than others – in particular, seeds, fruits, nuts and other propagules. That these are among the more robust parts of the anatomy of most plants is hardly surprising given that most have evolved as means of dispersing the parental genes. Other parts – leafs, twigs, bud-scales and thorns – can also be preserved. However, not everything is preserved; some plants may be important in the economy of a site, or a dominant element of the surrounding plant community, yet fail to be represented at all.

Macroscopic plant remains are used like pollen to indicate past plant communities and they can be used with pollen to distinguish taxa which grew close to the point of deposition, represented by pollen and macrofossils, and those which grew further away, represented only by pollen. They are useful in showing the presence of species which produce little pollen, such as cereal grasses, or which produce pollen susceptible to decay, such as poplar (*Populus* spp.). The nature of the deposit is obviously important, as the macrofossils in a detritus peat may have originated further from the point of deposition than those in a reedswamp peat. None the less, the combination of pollen and macrofossils has been used to great effect in elucidating long-term regional vegetation change.

Macrofossils are also valuable in smaller-scale studies of vegetational change and resource use. Charred plant remains, especially, are important because they are the main form of preservation in oxic deposits, and are recovered, often in quantity, from nearly all archaeological sites. Their interpretation, however, is not straightforward because of the complexities of the taphonomic trackways from live plant to deposition (Fig. 11.2). If plant tissues become incorporated within the embers of a fire, such that combustion occurs with minimal oxygen, then they may char, that is be converted to a complex and

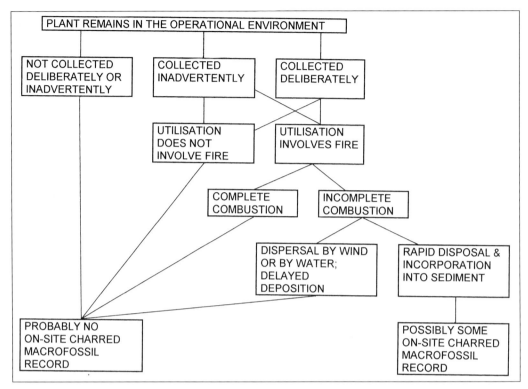

Figure 11.2 Taphonomic steps in the preservation of plant macrofossils by charring.

largely inert mix of carbon and other charring products. Charred structures are thus extremely stable, although vulnerable to mechanical damage. On excavations where burial conditions do not lead to preservation by waterlogging, charred plant remains may be the principal source of information about the local flora and, particularly, its utilisation. Plants which have been utilised in such a way as to bring them into proximity to fire are most likely to become preserved in this way. As a consequence, charred plant macrofossil assemblages have to be viewed in the context of human spatial behaviour and plant use and not as a random sample of the flora (Fig. 11.2).

In relatively arid parts of the world, preservation of plant remains by desiccation may also occur. This requires the plants to have been incorporated fairly rapidly into a forming deposit, thus avoiding destruction by detritus feeders or the destruction of refuse by fire or manuring, and then to have remained dry continuously to the point of excavation. These conditions are rare, but may be encountered in caves in arid regions, where they provide shelter from the relatively small amounts of precipitation. Desiccated macrofossils were among the main sources of evidence used in the pioneering study of early plant domestication in the Tehuacán Valley of southern Mexico (Byers 1967).

Interpretation of plant macrofossils is fraught with uncertainties, and the means of deposition has always to be kept in mind. The seeds and fruits recovered from refuse deposits at a small occupation site include taxa which formed the weed community of that site, perhaps attracted by the disturbed, base-rich soils, taxa which were utilised, and taxa which arrived 'by accident', such as in the droppings of synanthropic birds. Food plants may be obvious because large numbers of macrofossils of one taxon may occur together in a single deposit, implying human collection. Not all food plants are susceptible to this form of representation: a vegetable grown for its leaves may be abundant around a settlement, but seeds will be rare. The ruderal weeds of the site may indicate ground conditions. The interpretation of the macrofossils from a site may therefore focus on their value in indicating the surrounding flora and, from that, the environments around a site (e.g. Hall and Kenward 1990), or it may focus on human selection of resources and their uses. For instance, Pearsall's (1988; 1989) studies of charred macrofossils from Panaulauca Cave, Peru, led her to model the routes by which different species entered the site and by which charring took place.

WOOD

Trees may be encountered *in situ* as stumps within peat sequences, or as wood utilised and incorporated within an occupation site. In the former case, the environmental implications may be obvious. The common occurrence of tree stumps at the base of blanket peat and as submerged 'forests' in intertidal areas is evocative of major environmental change, which can be dated by dendrochronology (Baillie 1995). Long sequences of tree-ring width variation have been established for parts of Europe and North America, and sections of an ancient ring-porous tree such as oak (*Quercus* spp.) may provide an estimate of the date of death of the tree, and thus of the initiation of peat formation. In other parts of the world, tree-ring chronologies may be short and floating, but still of regional value, as in the chronology for Tierra del Fuego (Roig *et al.* 1996).

Ring-width variation may also reflect periods of more or less stress during the life of the tree. Periods of growth stress noted within one tree may correlate with stress seen in other trees of similar age, and thence to some substantial and widespread environmental factor. Periods of poor growth in northern British and Irish bog oaks have been correlated with an eruption of the Icelandic volcano Hekla in about 1159 BC, which is believed to have produced an ash and vapour cloud sufficient to disrupt weather patterns over a large area. Another period of narrow ring-widths has been noted for AD 536/7. This cannot at present be equated with a specific volcanic eruption, although other phenomena historically recorded around the same time point to marked weather excursions, crop failures and atmospheric phenomena consistent with a substantial atmospheric dust veil (Baillie 1994). The sensitivity of trees to different forms of stress can be tested by observation on modern trees. Sczeicz and MacDonald (1996) have used ring-width variation in white spruce (*Picea glauca*) to deduce variation in precipitation in north-west Canada over a 930-year period.

The utilisation of wood as fuel or for buildings or smaller artefacts brings it onto archaeological sites. To some extent, the wood identified among the charcoal fragments from an ancient hearth will indicate the species composition of the surrounding flora, but some assumptions will have to be made about the selectivity with which different timbers may have been used. The relative abundance of a particular taxon in an assemblage from a site may be no guide at all to its abundance within the woodland community. Equally, if selection has been practised, then the spatial resolution of the data is unclear: a particular timber may have been in such demand as to justify importation from great distances, and the nature of the demand may have been determined as much by cultural perceptions as by function. One important use of small wood such as twigs is as material for radiocarbon dating. As a twig represents only a few years' growth, the date of deposition of the twig is likely to be close to the date of its death, which is what radiocarbon dating measures.

Invertebrates

Invertebrates such as insects and snails typically have short life-cycles and may be highly adapted to a particular habitat or niche. They are therefore widely used in palaeontology as a means of reconstructing the environmental conditions in which a particular rock formation was deposited, and this approach has been adopted within archaeology. In relatively slow-forming deposits, usually not of direct human production, the corpses of the invertebrate community living on, in, and around the deposit will accumulate and may be preserved. The remains of arthropods, particularly beetles, and molluscs are amongst the more durable, and the use of invertebrates in environmental reconstruction has concentrated on these two groups.

Beetles

Beetles are one of the most diverse groups of macroscopic invertebrates, occupying every global biome except the sea and virtually every niche. Their exoskeletons have a complex chemistry, but, briefly, can be seen as composed of chitin, a complex polymer not dissimilar to cellulose in its chemistry. It is thus not surprising that beetle remains tend to be preserved in deposits in which plant macrofossils are also well preserved (Box 7.1). A dead beetle will disintegrate into the component parts of its exoskeleton, which include the head, thorax, elytra (wing cases), legs and abdominal sclerites. Of these various bits, it is the first three, and in particular the elytra, which are both robust enough to be commonly preserved and variable enough to permit identification to a high degree. Beetle fossils can, in the hands of one of the few experts in this field, mostly be identified to family, often to genus or even species. Remarkably, beetles have been evolutionarily stable: there is little evidence that evolution of morphology or the appearance of new species has occurred over many tens of millennia (Coope 1977). Beetles vary in their mobility, and whilst many taxa are relatively sedentary, and thus give relatively small-scale information, some swarm and fly considerable distances, particularly during the summer, and so may enter death assemblages at some remove from their breeding or feeding location.

The first stage in data collection is a list of taxa per sample, with numbers of specimens per taxon. However, a list of one hundred specimens may run to fifty or more taxa, most of them represented by a single specimen, and such data are difficult to comprehend, let alone interpret. One approach is through modern field studies which show that different habitats carry distinctive associations of species, thus allowing interpretation of fossil assemblages by analogy (Kenward 1978). As a part of this, taxa can be put into 'ecological groups', such as obligate aquatics or phytophages (plant-feeders), and it can be seen whether the majority of the sample consists of taxa from one of these groups. This approach assumes uniformity of adaptation on the part of different species over long periods of time, and that past habitats and environments will have modern analogues. Other approaches concentrate on the structure of the assemblage and quantify it by diversity indices or by other forms of numerical analysis which cluster the data into groups of high and low similarity (Perry *et al.* 1985) (p. 172). Whatever the approach, the aim is to interpret the assemblage in terms of the habitats and processes contributing beetle corpses to the deposit. In practice, this aim is generally attained from numerous samples in order to assess variation within a deposit, and thus, perhaps, spatial variation in the original ecosystem.

On an intra-site scale, beetles inform about ground surface conditions, vegetation, stored products and the utilisation of plant resources. Buckland *et al.* (1992) have used detailed sampling and interpretation to infer the use of different rooms within an excavated house in Iceland. At the other end of the spatial scale, the sensitivity of many beetle species to temperature renders them a valuable indicator of past climate. Morgan (1987) has used beetles to model early Holocene climate in North America, and Hoganson and Ashworth (1992) have monitored subtle climate changes in late Pleistocene deposits in Chile by examining changes in species and diversity in beetle remains in lacustrine silts. An excellent overview is given by papers in Ashworth *et al.* (1997).

Molluscs

Molluscs occur in the archaeological record as their shells. Being composed of calcium carbonate, they are only well preserved in alkaline deposits (Box 7.1), just the conditions in which insect remains are most readily destroyed, so there is complementarity between the two groups. Like beetles, snails include phytophages, detritivores and carnivores; they have generation times and rates of population turnover of a similar order to beetles; and they are found in often distinctive species associations which reflect particular habitats, thus making them potentially valuable in palaeoenvironmental reconstruction. On the other hand, the greater mobility of beetles has already been mentioned (p. 140), and snails are mostly less highly specialised than beetles, with few species occupying narrow niches.

The use of land snails as palaeoenvironmental indicators can be traced to studies by geologists of late Tertiary sediments, most particularly the work of Alfred Kennard (1943). From the 1960s, Sparks and Kerney developed the interpretative methodology in order to investigate land-snail assemblages from late Pleistocene and Holocene sediments (Kerney 1977), and Evans (1972) brought the study of snails into an explicitly archaeological context. Although snails are still important as a source of information on long-term, regional environmental change (Kerney *et al.* 1980; Evans *et al.* 1993), they are increasingly used to examine habitats and depositional circumstances within and around human occupation sites (Thomas 1985; Whittle *et al.* 1993).

Archaeological land-snail assemblages are often presented as a time-series histogram (Fig. 15.4) showing changes in abundance for ecological groups that are conventionally defined in terms of their need for more or less closed, damp ground conditions. A scale is

ENVIRONMENTAL ARCHAEOLOGY

Box 11.1 Stellmoor, Germany: integrating freshwater molluscs and ostracods

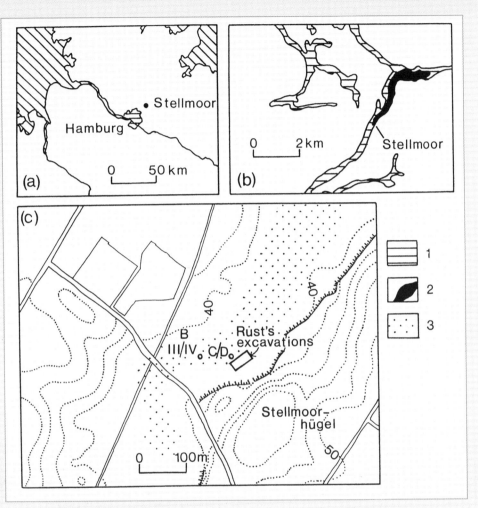

Location of (a) Stellmoor, (b) the tunnel valleys, and (c) the cores. 1 = tunnel valleys; 2 = Late-glacial lake; 3 = dead ice.

Stellmoor is near Ahrensburg, north Germany, and is one of many locations in northern Europe where the sediments of Late-glacial lakes can be sampled. The Stellmoor lake occupied part of a 'tunnel valley', originally produced by the action of subglacial meltwater (cf. Fig. 4.2). The site became well known in the 1930s, when excavations uncovered two Upper Palaeolithic horizons, including a wide range of artefacts, and bones of reindeer. The lake sediments were sampled to study the environment of the lake through time and the sedimentary processes which led to its infilling. This was achieved in part through examination of the sediments themselves, but largely through studies of the contained freshwater molluscs and ostracods. The sediments consisted of over 4 m of mostly calcareous marls and gyttja (organic mud), overlain by peat (Griffiths et al. 1994).

The lowest sediments indicate the lake to have had some through-flow of water. The sediments are relatively coarse, and the sparse freshwater molluscs are consistent with lotic (i.e. moving-water) conditions. Comparatively stable, biotically equable conditions quickly became established, with

BIOLOGICAL INDICATORS

Stellmoor, Germany: integrating freshwater molluscs and ostracods

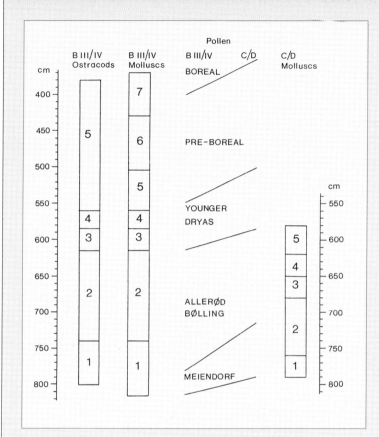

Correlation of cores BIII/IV and C/D with the pollen sequence; note that the numbering of the molluscan zones in the two cores is not equivalent.

diverse mollusc and ostracod faunas developing. The ostracods, in particular, include a number of taxa thought to be colonising species. A core taken towards the lake margin shows a gradual increase in phytophilous molluscs, but this is not reflected in a core taken from further into the lake basin. This suggests that vegetation was becoming established around the margins of the lake, though with open water persisting further away from the shore, and this may have been the setting of the older of the two phases of human occupation. This phase ends with evidence that the lake was becoming shallower. The sediments become more sandy, and the ostracods show some selective removal or destruction of juvenile forms, indicating that the death assemblage has been subject to wave action. A non-calcareous horizon, virtually devoid of molluscs or ostracods, follows and probably represents the beginning of the Younger Dryas cold climate stage. Subsequently, the lake re-established, and went through another stage of shallowing, with reworking of shells.

The importance of the Stellmoor study lies in the drawing together of sedimentary and biological evidence, to investigate questions particular to lake-edge and deeper-water locations, questions pertaining to the formation and history of the lake as a whole, and questions related to late Pleistocene climate fluctuations across north-western Europe. The biological evidence was analysed in terms of the presence or absence of particular species, their relative abundance, and the life-to-death translation. Within the freshwater molluscs, two cold-climate ecomorphs of more widely distributed species were noted: *Pisidium obtusale lapponicum* and *P. henslowanum inappendiculata*. In all, the project serves as a good example of the use of lithostratigraphic and biological evidence to address a wide range of questions on different scales.

drawn between taxa typical of, and normally restricted to, woodland leaf litter at one extreme, and those normally restricted to dry grassland at the other. This is useful in studies of Holocene woodland clearance, but less useful when examining assemblages from occupation sites or from very different environments. In particular, many assemblages may contain few individuals attributable to either extreme of this scale, and be dominated by the so-called 'catholic' species, a group which includes species of restricted, intermediate requirements and eurytopes of wide tolerance. One problem is that relatively little neoecological work has been undertaken on land and freshwater molluscan communities, with heavy emphasis still being placed on sources such as Boycott (1934; 1936), whose observations lend themselves more to the recognition of 'indicator species', rather than to the understanding of multi-species communities. Some studies of modern snail communities have been rather disturbing from the archaeological point of view, such as Cameron and Morgan-Huws' (1975) demonstration of high spatial variation in grassland communities. Similarly, Rouse (in Whittle *et al.* 1993) has shown that replicated sampling of a single, spatially extensive deposit can produce assemblages which vary in diversity and species composition. Some advances in interpretation have been made in recent years, both in the numerical analysis of data (Bush 1988a; Davies 1998), and in developing taxonomic groupings from the data rather than imposing them from without (Evans 1991). Land-snail analysis has been of considerable value in archaeology, and will no doubt continue to be so, but currently stands in need of a review of methodology and detailed investigation of modern analogues.

Other invertebrates

Mites Mites are arachnids, related to spiders. They may be abundant in invertebrate communities in soils, leaf litter and other terrestrial habitats, and include predators, detritivores and mould feeders. They have a robust, chitinous exoskeleton and so may be preserved in deposits in which other arthropods are preserved. A little work has been undertaken on the interpretation of mite assemblages from archaeological deposits (Morales Muñiz and Sanz Bretón 1994). Some early work attempted to characterise assemblages on the basis of all of the identifiable mite taxa and their known preferred habitats. Latterly, it has been found to be more productive to focus on particular groups of mites, in particular the Oribatidae, for which there is more neoecological information (Schelvis 1987; 1992). This effectively screens a palaeoenvironmental signal from the general background noise.

Ostracods Ostracods, a group of small, mostly freshwater crustaceans (Fig. 5.2), may be recovered in quantities from muds and marls, often in association with mollusc shells (Griffiths *et al.* 1993; Löffler 1986). Ostracods have particular requirements of water conditions and substrate, but are most useful when examined in parallel with freshwater molluscs (Box 11.1). Ostracod carapace fossils can also be subdivided by age and sex, so with a large assemblage it may be possible to investigate the population dynamics of a particular species and the season of deposition.

Cladocera Another freshwater crustacean which is not infrequently encountered in freshwater sediments is the water flea (*Daphnia* spp.). Unlike ostracods, however, it is not the adult carapace which is normally recovered, but the ephippia, highly resistant 'egg-sacs' which are produced by stressed *Daphnia* populations as a means of surviving unfavourable conditions. The presence of quantities of *Daphnia* ephippia in a sample, therefore, may reflect water conditions which were becoming marginal for the survival of this taxon, perhaps because of drought or eutrophication.

Chironomids Ancient lake sediments may also contain abundant remains of chironomids. These are small flies, commonly called non-biting midges, whose larvae, although largely soft-bodied, have robust chitinous head-capsules which can be identified with a high degree of precision (Bryce 1962; Hoffman 1986). Like diatoms, chironomids can indicate changes in water conditions and so, indirectly, information about terrestrial environments within the lake catchment. The study of chironomid assemblages has hardly impinged upon archaeology *per se*, despite having been used in reconstructing the Holocene histories of northern European lakes (Sadler and Jones 1997).

Nematodes The ova of nematode worms have been recovered from deposits on occupation sites, and have been studied as a means of investigating the alimentary health of past human populations (Jones 1985). Human populations, like those of most animals, have their internal and external parasites, and a light parasite load of fleas, lice, worms and mites was probably the normal state of affairs for most humans in the past. The nematode worms which inhabit the human gut, such as whipworm (*Trichuris trichiura*) and mawworm (*Ascaris lubricoides*), rely on the ingestion of ova as a mechanism of transfer from one person to another, and so produce prodigious quantities of ova into the gut, whence they are shed with the faeces. The ova have to remain viable in a transition from the gut environment into an unpredictable outside world, thence to the mouth and stomach of another individual. They are thus protected by a thick and chemically highly resistant wall, and the characteristic forms of nematode ova can survive burial in all but the most aggressively oxic deposits. The ova may be recovered from individual human stools, themselves preserved by mineralisation or desiccation, thus giving information about the parasite load of one individual. They may also be recovered from sediments around settlement sites where faecal disposal into pits, with concomitant contamination of ground surfaces, has taken place. In these circumstances, the concentration of ova per gram has been taken as an index of the degree of faecal contamination of ground surfaces and living areas, though it is not possible to equate a given density of ova per gram with some quantified degree of squalor. This is one field where the investigation of modern analogues should perhaps not be encouraged!

Vertebrates

The remains of vertebrates, as bones, are often the most conspicuous animal fossils in ancient sediments. Bone has a framework composed of collagen, the long molecules of which are twisted into helical bundles, which are themselves then twisted into larger fibrils.

The collagen framework is stiffened by impregnation with a mineral, hydroxyapatite, to give bone the characteristics of being stiff without being brittle (Vincent 1982, 147–60). Crudely, the survival of hydroxyapatite is favoured in dry or alkaline burial conditions: that of collagen in dry or anoxic conditions. Generalisations about the preservation of bone are easy to make, and generally wrong. The degradation of this composite material is poorly understood, though the importance of micro-organisms in the degradation of the collagen part is becoming more apparent (Child 1995; Nicholson 1996). Given these uncertainties, it is perhaps safest to say that bone may be preserved in a wide range of burial conditions, damp acidic conditions perhaps being the one major exception.

In most human populations, there is likely to be a higher intensity and frequency of deliberate interaction with other vertebrates than with invertebrates, and so the remains of vertebrates in archaeological deposits will reflect patterns of human behaviour more than the wider environment (Davis 1987; Hesse and Wapnish 1985). The larger mammals and some large birds may have co-habited with a human population as domesticates, or been its prey, and so the range of species and their relative abundance are heavily modified by human selection. Even so, the choice of domestic livestock has a bearing on the environment: Holocene woodland clearance in northern Europe is associated with the introduction of herds of domesticated grazers and browsers, cattle, sheep and goats. The remains of domestic livestock are thus relevant to examination of past environments in so far as they represent one of the human activities which modified, and responded to, them.

Smaller mammals and non-domestic birds may be indicative of particular environments, although the question of spatial resolution has to be considered. In studies of prehistoric sites in southern England, Evans and Rouse (1990) have utilised records of rodents and insectivore bones from the same samples as snail assemblages, often showing a good correlation between the two sources of information. Brain and Brain (1977) used small mammal remains from a rock-shelter site in the Namib desert to address specific questions about rainfall pattern and dune advance. Similar work in North America has concentrated on understanding the structure of the biotic community from which the death assemblage was derived (Graham 1985), and this has led to the use of accelerator radiocarbon dating to demonstrate the contemporaneity of an apparently ecologically disharmonious assemblage (Stafford and Semken 1990). In this regard, work on small mammal remains in North America is more sophisticated than that in the UK, where detailed biogeographical studies are more typical (for example, Yalden 1983; 1995).

Some attempts have been made to use assemblages of bird bones to investigate the predominant habitat around a particular occupation site, by ascertaining whether the prey available to hunters were mostly wetland birds, for example, or species typical of woodland. Bond and O'Connor (1998) have drawn attention to the concentration of winter migrant and winter-flocking ducks and waders in bone assemblages from medieval towns in eastern England, and have suggested a seasonal exploitation of wetland areas. Generally, though, spatial resolution is poor. Even rodents are multilocational in comparison with snails, and a migratory bird may have a life range which extends over thousands of kilometres (Box 3.2). Furthermore, there is always the difficulty with prey species that they may have been traded, and so deposited at a point far removed from the place of death or capture. As indicators of past environments, therefore, vertebrates are of limited use, except as an important part of the human operational realm.

There are obvious dangers in setting out a summary of the use of biological indicators of past environments against a taxonomic framework. Each source gives a partial aspect of the past ecosystem, and the integration of as many sources as possible is urged. It is also important to grasp the concepts of different taxa having different perceived and operational environments from each other, and from co-existing and co-locational humans. To take a simple example, a community of snails living amongst a patch of hawthorn (*Crataegus* spp.) scrub will be responding to changes in the environment as they perceive it to be, and the environmental parameters which loom large in the life of a snail (humidity and food) are quite different from those predominant in the perceived environment of the hawthorn bushes (soil moisture and light). Changes in the snail community will thus reflect changes in a different environment from that represented by the vegetation, and not merely in the sense that the snails have occupied a different spatial domain. Both snails and hawthorn bushes react to environments different from that perceived by a human population living in the same place. Some components of a past biota responded to short-term events, such as the drying-up of a ditch, whilst others reacted only to longer-term events. By bringing together data from a range of different indicators, therefore, we can obtain a rather patchy record of the past geographical environment, within which were set the operational and perceived environments of the human population which must lie at the centre of any explicitly archaeological investigation. Our concern is thus not to examine the past ecology of spermatophyte plants by means of pollen, or the past ecology of beetles by way of their exoskeletons, but to examine the past ecology of a complex community of organisms which included human beings.

In the next chapter, we consider the contexts from which evidence pertaining to past environments and ecosystems may be retrieved. All that we have outlined here regarding the use of fossil biota as indicators of past environments must be further constrained and modified by the context in which they are found.

Twelve

CONTEXTS

Contexts are a link between techniques (chapters 9–11) and interpretation (Part IV). Conventionally a *context* is the sediment or soil and other aspects of the physical environment in which an object is located. For a cereal grain, its context could be a pit and its assemblage of other cereal grains, and these in turn could be seen in the context of other pit assemblages and so forth. But more widely, a context is anything that contains information on past human environments, and a cereal grain itself is as much a sedimentary context for ancient biomolecules (Fig. 3.9) as soil is for a seed. We are concerned with different sizes and kinds of context, from individual items like a bone, through archaeological features like pits, and sites like hillforts, to bigger areas like uplands. This gradation does not reflect an increase in complexity or scale of relevance, but is a convenient form of presentation.

MOLECULES, ISOTOPES AND INCREMENTS

A cereal grain or an animal bone is a context. Its morphology tells us what species, or even sub-species, it is, and that it was part of the human diet; its ecophenotype about the conditions in which it grew; its state – hulled, charred, fragmented – about its processing and use; and radiocarbon dating its age. Such materials are also contexts for the deposition of *biomolecules* like lipids, proteins and DNA. These can be typed immunologically, for example, to identify blood in pigments and food residues, even to species. Proteins and DNA indicate evolutionary relationships since they change over time. They can indicate particular strains of plant and animal, which is especially useful where two or more of these are morphologically close, as with wild *vs.* domestic forms; and they can indicate geographical origins and spread. In combination with linguistics, biomolecular data from modern human populations give clues to their origins and spread (e.g. Hagelberg 1996). At the atomic level, bones and seeds contain elements and stable isotopes which can indicate the environment and in some cases the locality in which the organism grew (Fig. 12.1). For humans they can tell us about diet and its sources and environment (Keegan 1989). The ratios of $^{13}C:^{12}C$ preserved in bone, for example, can reflect the proportions of wild plants, certain domesticated plants, and the ratio of marine or terrestrial meat in the diet. Strontium *vs.* calcium proportions in bone can also indicate the relative importances of plant *vs.* animal foods and marine *vs.* terrestrial meat (Aufderheide 1989).

Similar principles apply to inorganic materials. Sources of metal ores, stone artefacts

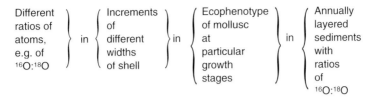

Figure 12.1 Different locations of oxygen isotopes and their relationship to shell increments, ecophenotypes and sediment layers (cf. Fig. 3.9).

and pottery are indicated by chemical, mineralogical (petrographic), physical or even geophysical properties. Movement of potential raw materials by glaciation complicates things because it spreads out formerly confined sources. With pottery, taphonomic problems can be sorted out by techniques such as mass spectrometry, e.g. the different sources of clay *vs.* filler, the location of kilns, and the contrasts between local, imported and imitation wares. These methods work because the most suitable properties of the raw materials for human use are usually narrowly localised and associated with equally localised sourcing properties.

Annual *growth increments* in coral, algal precipitates, mollusc shells and wood give data on annual and intra-annual environmental change and are especially useful for indicating season of collection in human food (Deith 1983); some marine littoral molluscs even show tidal increments. Oxygen isotope analysis of these layers gives greater precision about temperature, and in some corals up to ten separate readings per year can be obtained (Duplessy and Overpeck 1996). When such biological materials occur in contexts which themselves are annually layered, we have a further context for the same series of events (Fig. 12.1).

Oxygen isotopes preserved in calcareous biological materials like shell or as sediments can give an indication of contemporary temperature. The situation is complex because air, water-surface and water-bottom temperatures differ, there are differences between lake catchments depending on the amount of detrital material being washed in, the fractionation of different isotopes can have a species-specific component, and there are seasonal differences. Therefore, if the aim is to detail past climate, as is usually the case, analysis is best done on individual species, parts of mollusc shell or coral which were laid down at a known season or time of the year (e.g. von Grafenstein *et al.* 1992), or life stages which are active at, and contain the isotope signal of, one season only, as with some ostracod moults. If however the aim is to identify the local environment, we have a marvellous tool for examining the temperature diversity in one basin. And if we wish to identify the season of death or of collection, where shells are from a human occupation, we can measure the isotopic signature of the final increment of shell.

ARCHAEOLOGICAL FEATURES AND SITES

Buried soils

Buried soils occur as isolated patches in association with archaeological sites (Fig. 12.2) or as continuous distributions under deposits (Box 10.1). They contain long histories in a

ENVIRONMENTAL ARCHAEOLOGY

Figure 12.2 Bank and ditch sequence from the South Street long barrow, Wiltshire, showing the location of buried soils and sediments. Left: full transect; top right: ditch; bottom right: bank.

small thickness, so investigation needs to be intensive, especially as soils are often the only information about pre-deposit histories (p. 122). At the surface, horizontal variation can be considerable (e.g. Whittle *et al.* 1993), so horizontal sampling must be close (Fig. 7.6). Assemblages can be virtually autochthonous so samples must also be thin in order to minimise time-averaging. Close vertical sampling is also needed (Fig. 10.1) because of conflation of events, especially towards the bottom of the profile. Within soils there can be stratification, as with cultivation layers (Fig. 12.2) or inwash, and these can be identified by mineral magnetics, loss-on-ignition and micromorphology. Underneath the main profile there can be tree hollows, often separated from the main profile by a hiatus. These are valuable sources about earlier soil history (p. 125 and Fig. 10.5). Soils are often formed in materials which themselves contain fossil biota (Fig. 12.2) and it is vital to ensure that fossils derived from the parent material are separated from those intrinsic to the soil itself (cf. Fig. 15.4).

Ditches, pits and wells

In any depression there is often a basic quadripartite sequence of deposits (Fig. 12.2) (Cornwall 1958; Evans 1972; 1990). The primary fill forms by mostly physical weathering in a few years. In temperate climates, it is coarse gravel or scree and includes material from the pre-feature surface and subsoil. A snail death assemblage at this level can include shells from a variety of environmental and age contexts (Fig. 7.3), although the biota often reflect the specialised environment of the ditch bottom and primary fill (Evans and Jones 1973). Experimental earthworks (p. 187) have aided the understanding of primary fills as has an appraisal of the contexts themselves (Bell 1990). The idea that the sequences of fallen turves and rubble are annual may need re-assessing (Bell et al. 1996). The secondary fill also forms by physical weathering, but more slowly than the primary fill, and may also incorporate material from worm casts washed down slope. It is finer than the primary fill and towards the top is influenced by soil formation. The biota, although still reflecting damper and more sheltered conditions than the surrounds, are less extreme than in the primary fill and more representative of the site. The lowest part of the secondary fill forms within a few years of the hollow being dug and can therefore reflect the surrounds at the time of construction quite closely. At the top of the secondary fill, a soil forms which reflects the general situation of the site and its surrounds. If there is no further activity on the site, the soil remains active. However, many archaeological sites undergo further human activity and this is seen in the tertiary fill. This is often cultivation soil or colluvium, thus reflecting the wider locality around the site.

Depressions such as wells, ritual shafts and ditches on floodplains in which deposits were laid down in anoxic conditions, and which have remained anoxic to the present day, often contain organic materials (Box 7.1). These are especially important in dryland areas where such remains are rare. The Wilsford shaft in Wiltshire is about the only example of a collection of insects and uncarbonised plant remains from the chalk in Britain – outside river valley contexts – and is one of the richest collections of natural history from the British Bronze Age (wetland sites included) (Ashbee et al. 1989).

Occupation layers

This term, although widely used, is misleading. People live on surfaces, not amongst their refuse, and the consequence of people occupying a place, and walking across surfaces, is as likely to be erosion as deposition. Deposits which we call occupation layers are probably imported flooring, materials left in short abandonment events, primary rubbish, or secondary rubbish which has been deposited away from the actual area of occupation as middens, or even tertiary rubbish taken from the middens and spread on fields as manure (e.g. Davidson and Simpson 1984). All these can have concentrations of artefacts, ecofacts and organic material and stand out as dark layers from less rich deposits. Biologically, they are characterised by synanthropic species, and can occur in almost any situation. The term 'midden' is often applied to this sort of deposit, but Needham and Spence (1997, 80) suggest this be confined to refuse-rich deposits, with deliberate and sequential accumulation at one location. Suggested terms for other sorts of origins are: 'unitary dump' for materials dumped in a single act; 'undirected refuse aggregate' where aggregation is minor and incidental but in one place, as on areas of waste ground; and 'refuse-rich context' where no process can be demonstrated. These sorts of deposits are often deceptively simple, but they are some of the most complicated to study, not only because of the different kinds of deposit but because of the different origins of the materials within them and their likely re-use. Sampling needs to be fine enough to identify lateral and vertical distributions (Fig. 7.6, p. 87), and analysis needs to be done using macro- and micromorphology, and studies of modern formation processes by experiment and ethnoarchaeology (chapter 14). Interpretation in terms of human occupation itself must allow that the deposits represent refuse disposal, that they may therefore be spatially mixed, and that they are some distance from the actual occupation (e.g. Zeder 1991, 76).

A deposit which superficially resembles an occupation layer, yet which has a distinct and complex origin, is the 'dark earth' in urban sites of the late Roman and early medieval periods in the UK, in which charcoal, oyster shells, pottery and mortar are common (Macphail 1994). The dark colour derives from charcoal and humus. Micromorphology shows that the homogeneous, micro-aggregate structure is biologically reworked daub derived from plant-tempered plaster and clay from the walls and floors of collapsed and decayed timber-framed buildings mixed with occupation refuse – charcoal, ash and coprolites – by the biological activity of enchytraeid and lumbricid worms. Other processes involved are of directly human origin, such as where Roman abandonment left richly organic materials and burnt herbivore manure, or garden cultivation over burnt down timber and clay buildings. The main way these different origins can be identified is through micromorphology, but modern observations of soil formation in abandoned urban areas, as at bomb-sites in Berlin, have helped in providing a modern analogue (Sukopp *et al.* 1979).

Tells

Tells are discrete and visible settlement mounds, characteristic of middle and eastern Europe, the Balkans and south-west Asia. They are town sites in which successive building construction, dilapidation and repair – of mudbrick in the Near East, timber-framing with plaster and daub in central Europe and the Balkans – has led to their mound form (Rosen 1986). There is complex lateral and vertical variation. This is partly at a large scale, in that dilapidated surfaces are not regular and there can be pit digging and complete removal of some levels (Rosen 1986); and partly at a smaller scale especially with the successive, incremental renewal of floors and plaster walls, and sequences of levelled debris, occupation and dung-rich deposits in individual rooms and areas within these. Human activities from the scale of the individual – trampling, sweeping – to that of the society are represented, together with abandonment layers where more natural events are recorded.

The main contexts of a tell are:

1. The pre-tell environment: buried soils and deposits under the tell.

2. The off-site environment: on-site materials and off-site contexts like lake sediments and soils.

3. The repeated sequence through the tell of successive stages of construction, occupation, dilapidation (collapse and decay), buried soils and renewal, and the general sequence of these as an index of demographic expansion and decline (Rosen 1986). Mudbrick may contain snails which can indicate the origin of the material, but this is complicated by the presence of snails living on vegetation in abandonment stages (Bottema and Ottaway 1982). Short abandonment stages are represented by water erosion and the deposition of wind-blown materials, long ones by soil formation.

4. The tell buildings: individual buildings and their associations, including the different origins of the materials for different buildings (Rosen 1986).

5. Areas and space within and outside the buildings: micro-stratigraphy of floors (and walls), the impacts of activities on these, the deposits on them and post-depositional processes (Matthews *et al.* 1997). Patterns of rugs on the floors may even be recognised in the remaining dust (Matthews 1995).

6. Open areas and pits infilled with midden material.

7. Pedogenic and other geomorphological processes which affect the tell (Davidson 1976).

These contexts give us sequences of different lengths and frequency of recurrence: (a) the long and irregular ones of the pre-tell environment, the general trends through the tell (Davidson 1976), and the tell post-depositional environment; (b) the medium-term and perhaps more regular ones of the developments from building levels to abandonment, decay and renewal; and (c) the micro and incremental ones of activities in particular areas like reflooring and replastering. Spatially, activities can range from the scale of parts of individual rooms, to links with the off-site environment. Key studies include Asvan Kale, Turkey (French 1973), where on-site studies of crops and animals are integrated with off-site ethnoarchaeological, territorial and regional analyses; and Sitagroi, Greece (Davidson 1976; Renfrew *et al.* 1986), where sediment studies through the mound showed it to be derived almost entirely of building and occupation debris and not of wind-blown matter, where tell history and the alluvial sequence of the adjacent floodplain were linked, and where pollen analysis of a nearby lake provided a long vegetational sequence. At Abu Salabikh, Iraq, high-definition microstratigraphy and plant anatomical analysis have been applied to sequences of successive floors (Matthews *et al.* 1997).

Hillforts

Hillforts are similar to tells in their range of contexts, in the resolution of these (Fig. 12.3), and the fact that in both types of site there was probably much structured deposition, especially in pits (pp. 84–5). Soils beneath the ramparts, ditch fills, grain-storage pits (Fig. 7.4),

ENVIRONMENTAL ARCHAEOLOGY

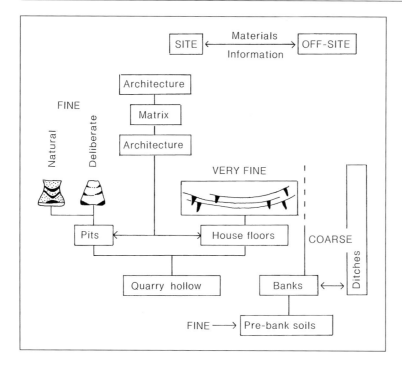

Figure 12.3 Sampling localities in an Iron Age hillfort. Contexts in approximate chronological order, although there is much overlap, especially of pits and house floors in the quarry hollows; the banks continue in use along with the ditches, although having less environmental information. 'Coarse', 'fine' and 'very fine' refer to potential sampling resolution.

quarry hollows used primarily to enlarge the inner rampart and then for house locations, the post- and stake-holes and superimposed floors of these (Fig. 12.4), and sometimes metalworking areas are the main contexts. Long sequences occur in the infills of the quarry hollows, including the house floors which overlie and are cut by pits. Some pits are infilled deliberately. Danebury (Cunliffe 1983) and Maiden Castle (Sharples 1991) in southern Britain have been investigated for the correlation of on- and off-site data, and the two areas can be directly compared because much of the evidence from both comes from similar contexts.

BIG AREAS

Uplands

Uplands are defined in relation to adjacent lowlands, and vary, therefore, with climate. In semi-arid regions, they are often areas of enhanced rainfall and thus of rich, often wooded vegetation by comparison with the steppe or semi-desert of the lowlands (Fig. 4.3). In temperate western Europe, they are often areas of higher rainfall and cooler climate, and thus of a shorter growing season. Lakes are characteristic, often originating from melted glaciers. Unenclosed grassland, heather and blanket peat are the main land types today (Figs 4.4 and 4.6), with the lower boundary around the upper limit of cereal cultivation, although this varies according to climate, demographic and socio-cultural factors. Solid geology is often hard, sometimes because of compaction by regional metamorphism, a part of the process of up-thrusting in the first place. This, with glacial scouring, has led to base-

Figure 12.4 A quarry hollow at Maiden Castle showing roundhouses, storage pits and post-holes.

poor soils with a tendency to leaching, iron-panning and, in high rainfall areas, blanket bog (Chambers 1993a). Grasslands are often used for summer grazing and breeding by domesticated and wild animals (Box 3.2) and subsidiary farming practices like cheese-making. In karst (i.e. montane hard limestone) areas, because of their inherent dryness, the best growth of vegetation and their use for grazing may be in winter (Fig. 8.2). Uneven topography and difficulties of communications impose restrictions on transportation, so that even if soils are fertile the areas may be perceived as marginal (pp. 189–91).

A microcosm of upland occurs along coasts where exposure limits productivity (Fig. 4.4). Although patchy and narrow, these areas are often unenclosed pasture, but with traces of past enclosure and settlement, and are thus an important record of lowland land-use. Offshore islands may present a particularly good record if there has been total abandonment by humans as social requirements changed and resources, especially fuel, were used up. Where combined with high rainfall, as in western Ireland, this coastal zone can be even more extensive, with blanket bog at sea-level.

ENVIRONMENTAL ARCHAEOLOGY

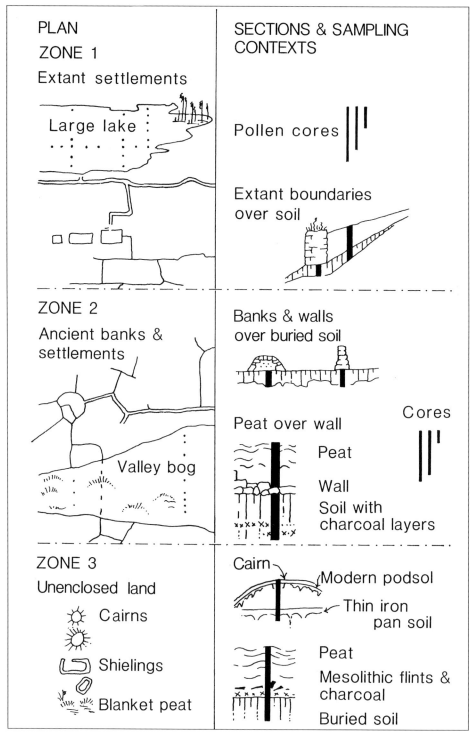

Figure 12.5 Upland archaeological contexts by zones, zone 1 being the lowest (see pp. 157–8). Black columns = sampling columns, dots = core locations.

Table 12.1. *Contexts for Bronze Age archaeology in an upland area, specifically the North York Moors, based on Spratt and Simmons (1976); units arranged in order of descending altitude*

Archaeology	Resources for environmental archaeology
High moorland and blanket peat; little used	Blanket peat; pre-peat soils; infilled lake basins
Watershed boundaries marked by round barrows	Edge of blanket peat; buried soils under round barrows; pollen and charcoal
Upland grazing, unenclosed	Modern soil profiles and soil pollen
Cairnfields on lower parts of uplands; ?summer settlements with shielings	Pollen analysis of buried soils under cairns and lynchets; ?ridge and furrow; localised valley and flush bogs
Steep valley sides with hollow-ways and other kinds of track	Surface survey
Lower-slope and valley-bottom settlements	Valley bogs and buried soils under earthworks; pollen

Typical contexts here are lake basins, raised and basin bogs, blanket peat and soils (Table 7.2; Caseldine 1990, figs 26 and 27). Upland in Britain is roughly zoned, according to altitude (Fig. 12.5), and it is helpful to examine the information potential of these zones in some detail.

Zone 1 is the uppermost zone of present-day enclosure and farms. Large lake basins provide a regional history and links with lowland land-use. Three-dimensional pollen analysis (Turner 1975) can identify general areas of farming around the lake (p. 64), while field survey can locate the farms and fields themselves (Table 7.2, p. 91). Dendrochronology dates artefacts and natural timbers in the lake sediments or bog. Stratigraphy, plant remains and insects, including those from bird pellets, reflect the lake and bog environments themselves. This is the context of our imaginary crannog in chapter 5. Sometimes, zone 1 merges uphill into zone 2, as in West Penwith, Cornwall, where ancient field boundaries are in use today and ancient settlements are overlain by modern ones (Johnson 1985). These are important contexts (p. 112). Often, however, there are bogs or steep unenclosed slopes with just trackways between zones 1 and 2 so that the two zones are quite distinct (Table 12.1) (Spratt and Simmons 1976).

Zone 2 (Fig. 12.5) is an unenclosed zone of upland grazing, usually at the limits of cultivation. There are ancient enclosures, cairnfields (clearance cairns) and farms, and their spatial definition is extensive. There are small valley bogs, or topogenous mires, and tongues of blanket peat from higher up. Buried soils preserve pollen sequences of site and local history; charcoal layers indicate burning, perhaps from deliberate vegetation management, and alternate layers of soil and peat indicate erosion (Whittington 1983; Mills *et al.* 1994), detectable by mineral magnetics. Dimbleby (1962) examined buried and modern soils of Mesolithic to present-day date from contexts such as these and tracked soil changes which took place as vegetation was successively cleared over long periods of time, even though each soil profile was only a short segment. Small mires immediately

Figure 12.6 A small upland topogenous mire in Cumbria, suitable for pollen analysis of local land-use history but unsuitable for a regional picture (see Box 6.1).

adjacent to earthworks can give a history of the locality (Fig. 12.6), directly related to settlement (Chambers and Price 1988; Molloy and O'Connell 1993).

Zone 3 (Fig. 12.5) is the highest land which has never been enclosed, mainly grassland and blanket peat. There are areas of Mesolithic flints and charcoal and buried Neolithic tombs under the peat (Fig. 8.3), and Bronze Age burial mounds on rocky high ground. Pollen analysis of blanket peat gives a regional picture of the vegetation, which can be compared with that from the large lake basins in zone 1. Pollen analysis at 1 mm intervals through small peat bogs around springs, with layers of charcoal, Mesolithic flints and, sometimes, mineral inwash has isolated several episodes of woodland recession, probably by burning, with a resolution of as little as ten years (Simmons *et al.* 1990). On watershed plateaux, lakes are important not only for increasing the range of environments but because they can contain calcareous deposits, with possibilities for molluscan and ostracod analysis, and bone preservation. The earliest deposits are usually of Late-glacial and early Holocene age, so there is the possibility of Upper Palaeolithic and Mesolithic data (Smith and Cloutman 1988). Later deposits may contain inwash layers providing correlation through pollen analysis and river terraces of farming histories and erosion (Fig. 12.7) (Tipping 1992; Mercer and Tipping 1994).

The contrasts which we draw here between land-use in upland areas and adjacent lowlands are a world-wide phenomenon. The mountain chains of Europe from Norway to

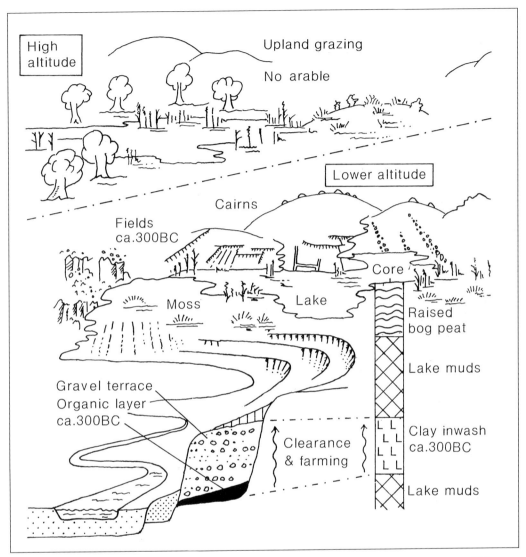

Figure 12.7 Correlation of upland agriculture, lake sediments and river terraces. The upper area is at high altitude where there was no arable and accordingly no signal in the lake sediments (not shown); the lower area is at a lower altitude where a period of land enclosure and arable c. 300 BC led to lynchet formation at field edges, inwash of clay into the lake, and the formation of a river terrace. (Loosely based on Tipping 1992.)

the Caucasus, the Atlas Mountains and the Andes all see the transhumant use of land by wild and domestic mammals and birds. Equally, some contexts are universal, especially lakes, acidic soils and the specialised and often highly visible archaeological sites. Blanket peat, on the other hand, although widespread globally, is very restricted in its distribution (Chambers 1993b, fig. 21.3), occurring in extreme geographical regions of high rainfall and cool climate.

River valleys

Apart from some continental interiors, especially of the African and Australian deserts, there are virtually no humanly inhabited areas of the globe in which topography does not relate to rivers (Brown 1997). Even in level plateaus of deeply weathered soils (e.g. Fig. 4.5) or in flat coastal marshes where sedimentation is estuarine, the water-table relates to river-valley topography. Likewise, karst areas even though characterised by solution, underground rivers and little surface-water runoff still have a general topography relating to surface river valleys. Rivers can cut across climate and regional zones, laying down sediments that reflect events further up the catchment, and creating environments different from those of the region in which the river flows, as in the rivers of the early civilisations. The main contexts are the valley bottom and the fragments of former river deposits preserved as terraces on the valley sides. In terms of processes, however, the entire catchment – its climate, the length of the river, gradient, relation to sea-level, valley form, soils, vegetation and land-use – is relevant to alluvial characteristics.

Basic contexts The present-day surface includes terraces at different altitudes (Fig. 12.8; Box 10.2), gravel fans coming in from the valley sides and active river channels. Archaeology occurs on the floodplain, often relating specifically to its management, as with water-meadows and flood embankments, preserved because of the absence of later settlement or cultivation. It also occurs on terraces, often only as cropmarks or surface scatters in arable land (chapter 9). Specific land zones which relate to river environments often have names referring to their use (p. 109). Deposits occur in four main contexts (Fig. 12.8): overbank alluviation on the floodplain; river channels; backswamp depressions; and valley-sides, where fan or outwash gravels and colluvium accumulate from sources immediately above the river.

Overbank alluvium, generally a minerogenic silty loam, is often the latest deposit and most valley bottoms and many terrace surfaces have at least a thin veneer. There are

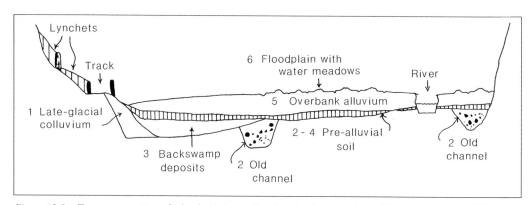

Figure 12.8 Transverse section of a lowland river valley showing Late-glacial and Holocene contexts. The numbers indicate a likely sequence, although the lynchets cannot be fitted in to this and the pre-alluvial soil could be contemporary or earlier than the backswamp and old channels where not cut by them.

variations in particle size, especially an upward fining which reflects the declining energy of flooding, and there can be buried surfaces within it. Overbank alluvium is primarily derived from the slopes of the catchment, reflecting land erosion further up valley (Fig. 10.8). In calcareous areas, a biogenic precipitate of calcium carbonate (Limbrey 1992) may form, more akin to lake marl or tufa than clastic alluvium. All these materials are suitable for fine analysis through molluscs, plant remains and micromorphology because they are often laid down as thin layers each year, not killing but rather enhancing the floodplain vegetation and leaving autochthonous assemblages (pp. 82–4). Remains of human activities such as manuring debris, charcoal layers, burnt mounds and enclosures are common. Beneath this material there should be a soil, developed into earlier overbank alluvium or Pleistocene gravel. This is often missed or noted as 'clay', but it is vital that it is analysed because it provides information about the pre-alluvial terrestrial environment. *Palaeo-channels* can be located by augering, aerial photography, or from boundaries on maps which reflect their meandering course. They often occur at the very edge of the valley bottom. The deposits are often mixed, with material from different areas and ages (p. 164). However the biota indicate the general environment of the catchment, which will be of a greater area than the ambient overbank alluvium, and there are often timbers suitable for dendrochronology and radiocarbon dating (Schirmer 1988). In the English Midlands, palaeo-channels, because of their small size, have provided pollen sequences relevant to habitations on terrace islands (Brown and Keough 1992), a similar relationship to that between small valley bogs and settlements in uplands (pp. 157–8). Sequences of palaeo-channels in central European rivers have been proposed in which the late Pleistocene pattern is anastomosed (channels linking obliquely across a weakly braided system) through a period in the Late-glacial when there were massive meanders, to a decreasing amplitude of meandering in the Holocene. In the Warta River, Poland, each type of palaeo-channel system is linked to a particular terrace (Kozarski *et al.* 1988). *Backswamp deposits* consisting largely of autochthonous reed peat and tufa generally occur in depressions towards the valley edge and are often of early to mid-Holocene age. This sort of deposit is thus valuable not just because it is another facies and environment but because it can take the local sequence back further than the overbank alluvium. *Valley-side deposits* allow us to identify relationships between alluvial history and the wider environment (Brown 1992). In periglacial environments material comes straight down slope, as with glacial outwash and fan gravels and Late-glacial material in English chalkland valleys (Fig. 12.9). However, when humans modify the land, valley edges are often complex, with field boundaries and tracks acting as barriers to the access of eroded soil into the river, so there is not always a signal in the valley bottom deposits of valley-side activity. Much colluvial material is also locked up in dry valleys (Fig. 12.9) which grade upwards from the river valley floor (Allen 1992). Three-dimensional recording of pottery allows the dating of the layers and the erosion episodes that formed them, which can then be correlated with adjacent field lynchets and settlements on the slopes and interfluves (Bell 1983; Smith *et al.* 1997).

Diversity in Holocene contexts Terraces form in areas of arid or semi-arid climate where erosion and downcutting are prominent due to a combination of patchy soils, patchy

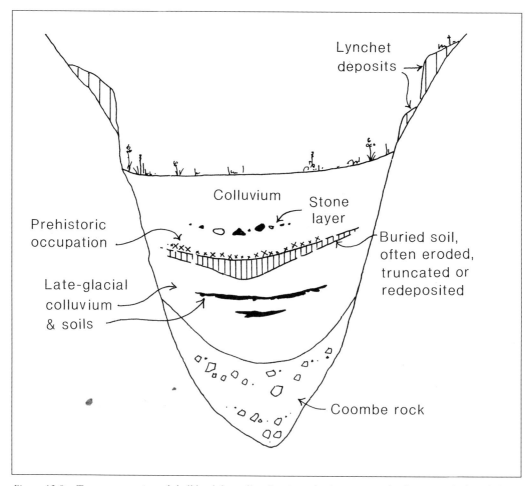

Figure 12.9 Transverse section of chalkland dry valley showing a basic sequence of colluvium (which includes coombe rock and lynchet deposits) and soils (cf. Box 10.1).

vegetation, strongly seasonal rainfall and high-rainfall events, e.g. the American Midwest (Box 10.2). Isostatic uplift of land also causes rapid downcutting and concomitant terrace formation, terraces grading downstream into raised sea-beaches. In uplands generally, because of steep gradients and rapid flow, terraces are common (Fig. 12.7). Downcutting caused by shifting channel and increased flow rate can also be of climatic origin (Needham and Macklin 1992). Individual terraces should contain the basic contexts outlined above, but they are often so fragmentary, and sometimes formed by lateral rather than vertical accretion, that the basal buried soil is missing. They often incorporate organic layers, with pollen and wood, so dated environmental histories can be obtained for each one (Schirmer 1988) and then put together as a sequence for the series as a whole (Fig. 12.10).

Rivers lacking Holocene terraces (e.g. Fig. 12.5) tend to occur in temperate climate areas with relatively thick soil, continuous vegetation (except where cultivated) and

CONTEXTS

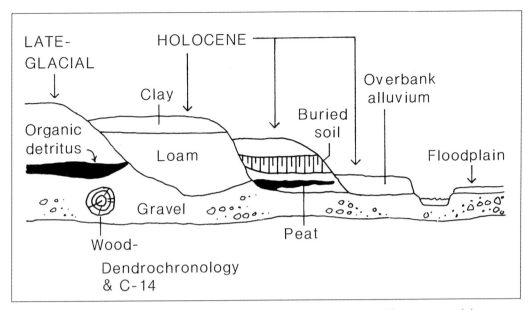

Figure 12.10 Transverse section of an upland river valley showing a succession of four terraces and the floodplain, with Late-glacial and Holocene depositional contexts.

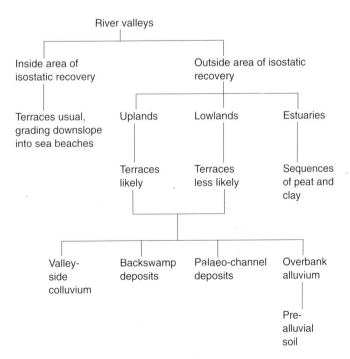

Figure 12.11 Classification of river valley contexts for the Holocene.

163

moderate rainfall all year round. They are also more a feature of lowlands, with gentle gradients as in many of the rivers of southern England where long continuous sequences through the Holocene can be obtained (Burrin and Scaife 1984). A basic classification is given in Fig. 12.11.

Cross-valley investigations are attractive, but they are only a part of the story. Contexts change from source to mouth. This is seen in the River Thames and its tributary the Kennet, as exemplified by studies by Evans *et al.* (1993) for the upper part of the Kennet where it is seasonally intermittent and where there are deep deposits of late prehistoric alluvium; by Wymer (1962) for the middle reaches, including the important Mesolithic site of Thatcham; by Butterworth and Lobb (1992) for its confluence with the Thames, where there are Bronze Age deposits in channels, overlain by peaty soils of the Roman period and then medieval alluvium; by Robinson and Lambrick (1984) and Limbrey and Robinson (1988) for the Upper Thames where a soil profile on gravel is overlain by alluvium of Iron Age and later date; by Needham (1991) for the Neolithic and later deposits at Runnymede in the Middle Thames, an important site for its diversity of data; and by Devoy (1979) for the Lower Thames where there are deep intercalations of fen peat and inter-tidal clay.

Pleistocene contexts River valleys take us back into the Pleistocene, with terrace sequences from south-east England, western Europe, across into Georgia, southern India and as far east as Java having long hominid histories. Pleistocene terraces are related to global changes of sea-level (eustasy) and are on a grander scale than Holocene ones. In the Thames and the Somme, they extend to nearly 270 m above present sea-level and grade into sea-beaches of equivalent age. Deposits are often thick and extensive as they were often laid down during periods when glaciers were melting and releasing huge quantities of water, with rivers moving across the floodplains in a braided or anastomosed, rather than meandering, pattern. In some cases there are palaeo-channels beneath present river levels and, where there is downwarping of the land, these grade down below present sea-level. Within the gravels are layers of finer loams (brickearth), sands, organic layers and, reflecting periods of river recession, soils and loess. It is these, rather than the gravels as such, that contain Pleistocene archaeology and biological materials, often on a gravel surface covered by fine alluvium (Wymer 1992).

River caves In limestone areas, rivers are largely underground. When the channels dry up and the land is eroded, these can appear at the surface in cliff faces as caves. They have some of the longest Palaeolithic and hominid histories. Some sites are relatively simple, with the cave being a part of a former river, exposed by valley-side erosion and enlarged by frost weathering. Deposits consist largely of thermoclastic rubble with soil horizons reflecting warm periods, and they usually occupy two contexts, the cave proper and the talus slope outside the mouth of the cave. Sequences in the Rhône delta area (de Lumley 1972) and the Perigord region of south-western France cover a period thought to extend from the Riss (Penultimate) Glaciation to the end of the Würm (Last) cold stage and include a detailed record of interglacial and interstadial stages (Bordes 1972; Laville *et al.* 1980), with occupation by animals and hominids. Other sites, such as many of the southern African early hominid sites, are more complicated, some starting life as solution hollows rather than rivers. Animals and geological processes incorporated bones into the deposits. The steep slopes of the sink-holes provided a refuge for trees out of reach of grazing animals and the trees in turn provided shelter for leopards to eat their prey away from scavengers. The bones of these meals, including hominids, fell into the hollows and were incorporated into travertine more durable than the surrounding limestone, thus remaining after erosion as hillocks (Brain 1981). These were mined for their high-quality rock, leading to the discovery of the hominid bones.

Coasts and estuaries

Coasts are the ultimate ecotone, the 'double larder' of land and sea (Brothwell and Dimbleby 1981), more intensively settled than any other zone of the globe. They are ever changing and have contexts of a greater variety and in closer juxtaposition than anywhere else. There is also high visibility of time in coasts, with fossil features like inter-tidal peats and 'forests', stabilised dunes, land-locked lagoons, raised beaches and redundant sea cliffs all attesting to a depth and complexity of history. Tectonic changes – land uplift, land downwarping or sea-bed movements – are widespread, there being few areas of the globe that are utterly stable. In coastal areas, sedimentation or shoreline elevation are the two main results of subsidence and uplift respectively, providing a range of contexts. Glacial isostasy (the depression and uplift of land associated with glaciation) is one of the main causes of coastal change.

In subsiding areas, there is often a patterned distribution of deposits (Fig. 12.12) especially where prevailing winds are onshore. In estuaries where land is flooded, there are

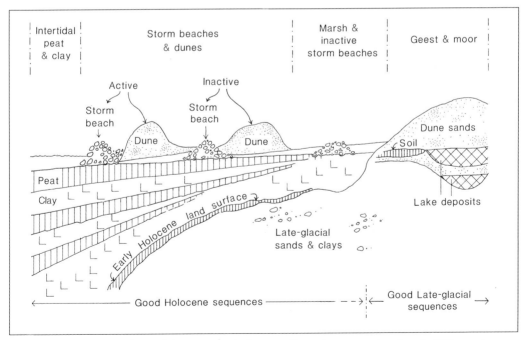

Figure 12.12   *Sequence of coastal deposits outside the area of isostatic recovery, based on the Dithmarschen area, Schleswig Holstein.*

clays which support saltmarsh. In more exposed areas, storm boulder beaches grade landward into active sand dunes. To the landward there are fossil beach ridges and older, stabilised dunes which in turn give way to fixed-dune pasture, extensive areas of grasslands (and sometimes arable) called 'machair' in the southern islands of the Outer Hebrides (Gilbertson et al. 1996). These sands are often calcareous and fertile with, accordingly, plenty of evidence of former occupation, middens and cultivation. Often, settlements which are being eroded by the sea were once at the edge of inland lakes, hundreds of metres from the coast (e.g. see Box 8.1). This is seen in freshwater lake deposits now in the intertidal zone, often intercalated with peat and tree stumps ('submerged forests').

Intertidal deposits are a visible part of a more substantial unit (Table 7.2, Fig. 12.12), and were formed in northern Europe during early Holocene global sea-level rise. Estuarine clays and peats extend landward beneath the storm beaches (German, *Nehrungen*) and blown sands, formed in a series of wetland successions, variously of reedswamp to carr woodland to raised bog. They appear at the surface as extensive areas of marsh (German, *Marsch*) and reedswamp on which there are mounded settlements (known as *Wierde*), often grouped along the slightly raised land of the fossil storm beaches. The land is protected from sea erosion by dykes, some of the Middle Ages. Underneath these deposits is a fossil land surface of lakes, raised bogs and soils. Being buried beneath several tens of metres of deposits, these are difficult to study but they appear in commercial excavations for docks and large industrial plants when archaeology is often found. The land surface rests on Late-glacial sandy deposits formed by glacial outwash, and later by wind deposition from sand banks of rivers, like the Elbe and Eider. These extend inland and upwards, outcropping as extensive sheets and fossil dunes, or coversands (Fig. 4.2), which back the lands of marsh, ridges and settlements. The sands are non-calcareous, their soils often podsolic, and today they support heath and coniferous woodland. This is termed *Geest*. The sands may overlie calcareous freshwater lake sediments of the Late-glacial Interstadial which are valuable sources of Upper Palaeolithic material. In low-lying, wetter areas there is *Moor*. This coast to inland sequence of 'dune' through 'Marsch' to 'Geest' and 'Moor' is a pattern in north-western Europe, bordering the North Sea.

In areas of isostatic uplift, as in Denmark and northern Britain, ancient coastlines are preserved some distance inland, often from earlier in the Holocene and Late-glacial periods than elsewhere. Visible traces of settlement like the Mesolithic shell middens around the coasts of Denmark, south-eastern Scotland and the Inner Hebrides are widespread. The raised beaches grade into river valley terraces and, in peripheral areas of the isostatic uplift zone, into estuarine clays of the earlier Holocene sea, especially in long estuaries, as with the carse lands of southern Scotland.

Marine and non-marine molluscs occurring together in the same contexts allow a study of associated land and marine changes. Land molluscs indicate the prevailing vegetation and the varying influence of blown sand from the coast, while the marine molluscs, usually from middens, allow correlation with contemporary littoral changes – assuming they derive from close to the site (Evans 1972). Ecophenotypic variability in the marine shells can indicate degrees of storminess, while oxygen isotope and growth-line analyses can indicate the season of collection (Deith 1985; 1990) and climate (chapter 11). Where raised beaches grade downwards into estuarine clays, and these into lake deposits (Singh and Smith 1973), there are opportunities for studying changes in molluscan communities from freshwater, through brackish, to full marine.

Ice-sheets, as well as having caused tectonic changes, locked up huge quantities of water so that sea-level fell many tens of metres and coastlines were often many kilometres from their present positions. Melting during warm stages submerged much Palaeolithic and Mesolithic archaeology – perhaps the best, since the areas lost were largely grassy, well-watered and game-rich plains. Sea-level rise during warm stages was often to successively lower levels so that raised shorelines of each period occur as a series of steps, the higher ones older than the lower. The situation is the same as with river terraces (pp. 162–4), and indeed these shorelines often grade into the terraces in estuaries. In some cases, as at Boxgrove in Sussex (England), Lower Palaeolithic stone tools and animal and human bone occur on land surfaces just on top of the beach deposits. Sea-caves are also preserved as part of these ancient shorelines. A common sequence, shows Last Interglacial beach deposits overlain by cave earth and rubble, with thinner layers of travertine and soils. At Klasies River Mouth, in South Africa (Singer and Wymer 1982) the deposits contain Middle Stone Age archaeology and hominid remains from 120,000 to 60,000 BP.

Two factors have led to the popularity of many of these contexts for archaeologists: their strong visibility through erosion and the fact that they are on land which is marginal to present-day settlement and farming (Fig. 12.13). To this extent, research is reactive, and it should be broadened to include areas of hidden archaeology, especially at the interfaces between obvious remains and more cryptic ones which are equally areas that people used and settled because of the local diversity of soils and resources.

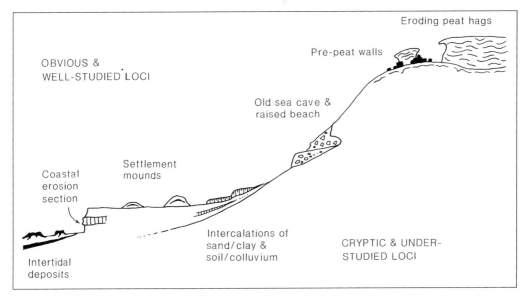

Figure 12.13   Transverse section from upland to coast showing the dichotomy between obvious and well-studied loci and cryptic and under-studied ones; loosely based on the southern Outer Hebrides.

ENVIRONMENTAL ARCHAEOLOGY

Figure 12.14 Dalmore, Lewis, Outer Hebrides. Coastal bay showing the interface between blown-sand and mineral soil. Note the modern settlements and agricultural ridges. Much archaeology takes place in the blown-sand areas where it is eroding, whereas there is probably a good deal to be located in the areas of the modern settlements as well (cf. Fig. 12.13). Photo: Historic Scotland.

Part Four

INTERPRETATION

Interpretation is going on all the time, from the influence of our own social and cultural lives, through the moment we decide to study past environments and the decision to do this at particular scales, to the selection of specific tracks and codes of entry into the interpretation itself and the final publication. It is necessary for data gathering in the first place (p. 93) and then, with regard to the data and reassessment of research and sampling policy, it becomes iterative and cumulative. It is also creative, especially with regard to humans (Shanks and Hodder 1995).

Interpretation is done at different levels along the chains of cause and effect (Fig. 10.7). *Proximal factors* are those to which the raw data relate closely, like temperature for an animal, water chemistry for soil, and cartography for maps, while *distal factors* are those which are more directly relevant to humans, like climate affecting the temperature regime, disturbance affecting soil water chemistry, and settlement history affecting the intensity of map-making (Fig. IV.1). Also the organisms, soils or documents themselves constitute part of the environment and can influence the proximal and distal factors. Distal factors can result in the same combination of proximal factors: the same map may be drawn for reasons of local economic change or regional climatic or environmental deterioration. Although there is a gradation between proximal and distal factors, they must be kept distinct conceptually when interpretation is going

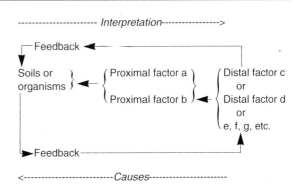

Figure IV.1 Relationship between soils/sediments/populations and the proximal and distal environmental factors to which their behaviour relates.

on. It is quite valid, and necessary, to interpret data immediately and intuitively in terms of distal factors, such as suggesting that a blip in a pollen diagram is due to humans burning vegetation, but we still have to be aware that there are steps to take between those data and this inference. There is always a range of distal factors for particular sets of primary data and there must be open-mindedness about this. Polarisation of debates, especially between the environmental impact of climate and people (e.g. Chambers 1993b), is damaging (p. 133).

We must also use a range of interpretational procedures, and accept that the extremes of inductivism and analogy are helpful and that there is no clear division between them. Inductive reasoning identifies patterns in raw data in relation to themselves and adjacent contexts without the input of external information (chapter 13). However, it uses data selected in an analogical framework – individuals, species, distributions – which seem objective, but are not, since we decide what framework to use. Equally, using inductivism in the first place implies that some patterns are significant, and that decision requires analogy. Within analogy there are gradations depending on the data, its known relationships with processes, and how far along the road of inference we choose to go. Thus analogy is relevant for the world of soil and sediment processes where there are clear rules into the past, less so for the biological kingdoms where there are varying ecologies and behaviours within each species, and even less so for the human sphere (chapter 14). *Relational analogy* is when we infer general aspects of the environment that we believe have always held, as with the 'pyramid of numbers' which connects trophic levels in an ecosystem (Part I). Such relationships may be easy to substantiate but the information obtained is general. *Formal analogy* using proximal factors means not going far along the chain of inference and again indicates general features of the environment, like disturbance. Formal analogy using distal factors is where data are matched closely and narrowly to environments with reference to the present day. This is not so reliable because a specific relationship between data and inference may not always have held, and because different factors may lead to the same end point (equifinality). A beetle species that requires bare soil, sandiness and warmth can find these in arable, partially cleared scrub, bare soil around farms, areas of overgrazing, flooded land where there are thin layers of dried sand, or in river channels on exposed sand bars.

Interpretation of physical environments often uses circumstantial evidence. Erosion episodes can be said to be causally related to human activities if there is a regional correlation in timing between them and, say, periods of demographic expansion (p. 212). But this is 'guilt by association' (Gould 1980). Lower Palaeolithic associations of large mammal bones and human artefacts do not mean hunting, although the recent discovery of yew 'spears' from Schöningen in Germany strengthens the case for this (Dennell 1997; Thieme 1997). The association of charcoal and vegetation clearance, pollen diagram blips and hunting artefacts does not mean the deliberate and controlled use of fire; and the association of artefacts, a pollen record of clearance and cultivation, human demographic peaks, and episodes of alluviation does not mean that alluviation was caused by agriculture (Fig. 12.7; Fig. 10.8, p. 130). We need unique links between past life processes and their signal in the archaeological record.

In chapter 13, we review some methods of data analysis which facilitate the recognition and testing of patterns in the data. Chapter 14 reviews the use of present-day observation as analogy, and in chapter 15 we use a series of case studies to draw together some of the points and ideas outlined in earlier chapters.

Thirteen

Data Analysis

A lot of interpretation is based on visual assessments of distributional correlations, say of soil types and archaeological sites, and time sequences, as with the zonation of pollen diagrams. But much information remains hidden, especially about past associations which may have no analogue at the present, and it is the aim of this chapter to show how this information can be revealed.

The steps are:

1. Simplification of the raw data. This is the establishment of characteristics for a data set which can be compared with others. A list of ostracods cannot be used as such, but a fifty-species list can be immediately compared with a two-species list and interpreted in terms of ecosystem complexity.

2. Pattern and group identification. This means comparing the internal structure of data and their association with contexts to identify patterns which may be mutual, complementary or random, and is based on the premise that human, biological and geomorphological systems impose patterning on land (Hodder and Orton 1979, 9) and through time. In a list of fifty species of ostracod, significant associations of three species in particular abundances, and with a particular sediment, can only be revealed from the identification of a repeated association. This, too, is the basis for the zonation of pollen diagrams or the identification of associations of particular archaeological sites with particular time or space zones. However, in these, the assemblage order or site distribution is fixed, and this means that groupings refer only to one particular case, and there is a tendency to emphasise blocks of adjacent assemblages or sites with closely similar characteristics and to ignore zones of change or isolated examples.

3. Feedback of these groupings to the raw data. This is the enhancement of a single data set with reference to groupings established by comparison of several data sets from different sites and areas; the significance of a particular assemblage or pattern is strengthened with reference to the whole (Evans and Williams 1991).

4. Statistical testing of the significance of groups thus established. Some patterns remain hidden to visual analysis, while some apparent patterns are insignificant (Hodder and Orton 1979, fig. 1.5). Mathematical methods for simplifying data, identifying patterns (or randomness) and making correlations, and statistical testing for their significance, including the acknowledgement of their contexts through Bayesian statistics (Buck *et al.* 1996), are an essential step (Shennan 1988).

GROUPING DATA INTERNALLY

Some methods group data on descriptive categories, such as species or soil types, basing these categories more on general principles of biology and ecology (Southwood 1978; Pielou 1975; Magurran 1988) than on particulars of species' ecology or detailed pedogenic processes.

Nearest neighbour analysis

This is a method for analysing spatial distributions and statistically identifying groupings or randomness of point data. It can be used at the site scale for analysing distributions of bone and artefacts, the locality scale for vegetational or animal distributions, or at the regional scale for settlements (Hodder and Orton 1979). However, frequent gaps in our data at the larger scale make them unsuitable for this form of analysis.

Spatial associations

Lateral spatial analysis can also be applied to two or more types of data such as species or assemblages. Hodder and Orton (1979, 200) give various measures. The sorts of relationships that are being identified here are whether two species occur independently of each other (no overlap in distribution), partially overlap, or are totally coincident. We can also look at species or archaeological distributions in relation to environmental factors like soils and topography to see if there is a relationship (Attwell and Fletcher 1987) (Box 14.4). Other forms of lateral spatial analysis have been used, for example by Greene and Lockyear (1994) to examine interspecies differences in the distribution of plant macrofossils across an excavated surface, and by Williamson (in Lilley et al. 1994, 371) to model the distribution of different age and sex classes of humans in unexcavated parts of a cemetery (compare the use of GIS, Table 9.1, no. 8).

Diversity and equitability

Data sets can be characterised without recourse to individual species' ecologies, the environmental implications of soil or sediment types, artefact uses, or settlement functions. With biological material this means numbers of individuals, biomass, age and sex structure, taxonomic affiliations and morphology (chapter 3). A simple way of doing this is with rank-order curves (Fig. 13.1). These are usually used with species assemblages but could equally be used with artefacts. Note that the tail-off in these diagrams may be due to residual individuals, not a true reflection of the former community. An extension of this is the use of diversity indices which incorporate inventory diversity and evenness (pp. 28–9) either separately or together (Pielou 1975; Magurran 1988). The Shannon–Wiener index, H', incorporates both, although only taking relative abundances (Figs 13.1 and 15.4). In Fig. 13.1, the environmental ascriptions are based on the ecology of the species. There is clear distinction between (a) and (c) in terms of rank order and H', supported by the species ecology; (b), however, is difficult to separate from (c) in terms of rank order and

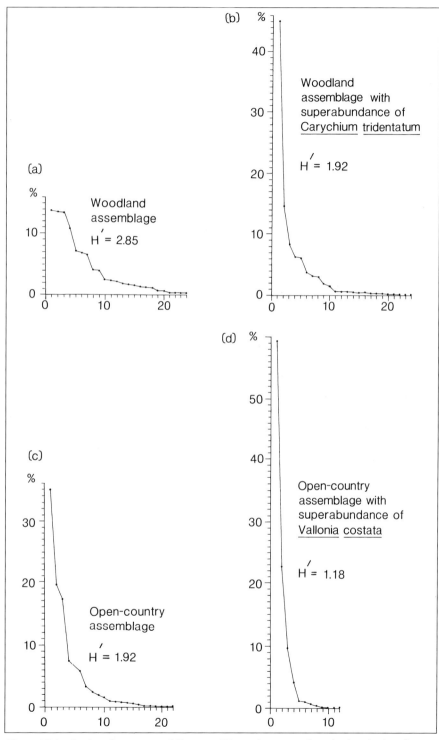

Figure 13.1 Rank order curves and Shannon–Wiener diversity indices for four land-mollusc assemblages.

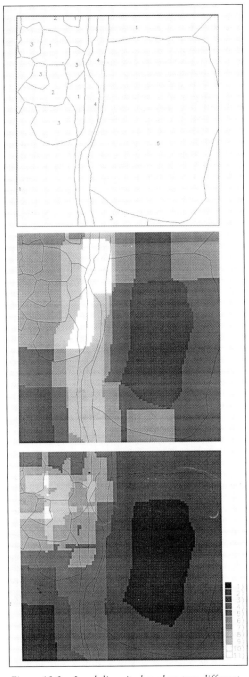

Figure 13.2 Land diversity based on two different calculations of the same area (top). Pale = high diversity, dark = low diversity. (Diagrams prepared by D. Wheatley.) See p. 175 for explanation.

H', yet on species-ecological grounds the two assemblages are very different; (d) is unequivocably an open-country assemblage on all grounds. Simpson's lambda (λ) is the probability that any two individuals picked at random will be the same species and is therefore a useful measure of how individuals in a sample are concentrated into a few species. High values indicate high bunching, so the index is an inverse correlate of H'. Fisher's alpha (α) has been extensively used by Kenward (1978), amongst others, as a combined measure of species richness and equitability in beetle samples. It differs from the Shannon–Wiener index or Simpson's lambda in being based on the number of species in the sample and the number of individuals, so it is similar to the Brillouin index. Use of these indices has been criticised, especially by neoecologists (Price 1975; Gee and Giller 1991), because they combine two components and because in fossil assemblages there may be different ecological groups through mixing at death, during deposition, or during sampling. However, it is the combination of different kinds of data in one index that makes them attractive. Comparisons can be made with inventory diversity and the Brillouin index to identify the contributions of different components, and the degree of mixing may be assessed by comparing H and H'. Moreover, no one would consider using diversity indexes for mixed faunas of aquatic and land species. The message is clear: rank order and diversity indices can be useful in some cases, but they must be used with other methods, and the taphonomic trajectory of the assemblage must be considered.

The concept of diversity can also be applied to land. Units of terrain may be defined in terms of one factor (e.g. soil type) or a combination of several (soil, altitude, and slope) (Fig. 13.2, top). Two attributes of diversity can be measured: (a) diversity in terms of potential access by humans (Fig. 13.2, middle), which takes into account access to the low diversity land of category 5; and (b) diversity as a measure of patchiness, i.e. distribution and size of the units (Fig. 13.2, bottom). This approach begins with a GIS model of the terrain (e.g. see Box 14.4), and then uses tailored software to compute the different diversities (David Wheatley, personal communication). Note that neither of these types of land diversity need be equivalent to species inventory diversity. The individual units in an area of high land diversity – a copse, fields, a settlement – may each have a moderate or low inventory diversity, whereas an area of low land diversity can have a high inventory diversity if it is a large mature woodland, or a low one if it is semi-desert.

The same approach can be applied to sequences of sediment units in terms of occurrence of different thicknesses of unit. A high diversity sequence is one with a rapid succession of different sediment types, each one of which is likely to have a low inventory diversity because of the rapidly changing environment, while a low diversity sequence where there is little change can have a low inventory diversity, as with loess or blanket peat, or a high one if it is tufa or coral.

Methods of assemblage comparison

We now consider the degree of difference or similarity between assemblages.

Similarity indices There are numerous indices of the degree of similarity of pairs of compatible samples of biota which use either numbers of individuals for each species or presence or absence of a species (Fig. 13.3). The caveat made above concerning the difference between a modern system of known spatial and temporal extent and a sample of an ancient community of uncertain precision remains. The choice of index depends on the questions being asked and the quality of the data. Results may depend on the index, sample size and species diversity. For example, Wolda (1981) studied similarity indices against simulated data for which sample size and diversity were varied. Diversity, and thus the relationship between species richness and number of individuals, were assumed to follow the log series distribution on which Fisher *et al.* (1943) based their diversity index alpha (α). The starting premise was that two random samples from the same 'fauna' should give a close approximation to the theoretical maximum value for that index, and Wolda concluded that maximum values of virtually all indices are influenced by sample size and diversity. Despite this, several indices have potential, providing large disparities of sample size and diversity are avoided. Indices which take account only of presence or absence include the Czekanowski (or Sorensen) coefficient and Mountford's coefficient. Both require knowing only the number of species in each of two samples and the number common to both. Some other binary coefficients require knowing the number of species absent from both samples, which is not really calculable for fossil material. Of the indices which use relative abundance to compare samples, one of the simplest and most robust is Renkonen's percentage similarity. Another is the Morisita Index which is independent of sample size and diversity – although presenting a major challenge to the less numerate! Algorithms for calculating these, and many other, indices are given by Wolda (1981).

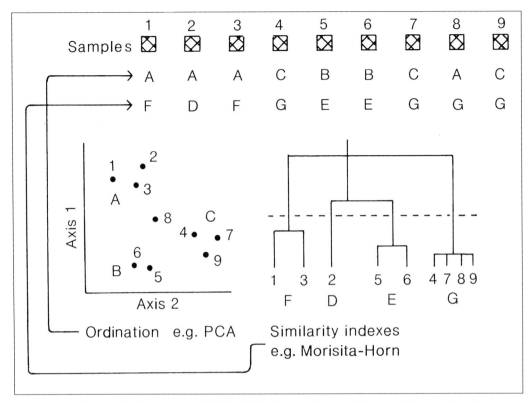

Figure 13.3 The use of ordination and similarity indices in grouping assemblages and characterising them by feedback.

Cluster analysis The results of similarity indices can be expressed in matrix or table form or, more usefully, as a dendrogram, the groupings in which can then be identified by cluster analysis (Fig. 13.3). At its simplest, this is a method of altering a matrix of similarity values so as to produce a dendrogram on which different degrees of dissimilarity between clusters of samples appear on different branches of the dendrogram. There are mathematical methods to determine whether the groupings are significant or part of a continuous series that has been artificially created by low sample size or just the seriate nature of the assemblages. Identifying which clusters are significant, i.e. where to draw the line between the single cluster of all the data or the individual assemblages, can be done by eye (the usual method) or mathematically (Gower 1975; Everitt 1980; Pielou 1984).

Cluster analysis has been widely used in archaeology as an adjunct to the typological analysis of objects, though the mathematical principles are the same as in analyses of fossil biota. Orton (1980, 46–54) gives a summary of the strengths and weaknesses of various forms of cluster analysis. A good example of its use in bioarchaeology is given by Benecke (1988) who, in a study of the animal bones from a medieval site at Menzlin, Germany, used cluster analysis to examine the degree of similarity of a series of broadly contemporaneous north German sites.

Ordination techniques These group data in terms of closeness, either of the way species behave in relation to each other or of overall assemblage composition (Fig. 13.3) (e.g. Thomas 1985). Commonly used methods are principal components analysis (PCA), detrended correspondence analysis (DECORANA) and two-way indicator species analysis (TWINSPSAN). The methods are useful for identifying which factors (axes) are important in separating assemblages and in confirming visual groupings (Bush 1988a; Rouse and Evans 1994; Cong and Ashworth 1997). They can also be used in morphometric analyses of biological material, as in the study of Benecke (1987) on the separation of wolves, wolves at an early stage of domestication, and dogs.

The groupings established by cluster analysis and ordination amplify the raw data (Fig. 13.3) and are characterised in terms of them, as by diversity and salient species groupings.

METHODS WHICH USE INDIVIDUAL SPECIES' ECOLOGY

Other methods of grouping data use ecological or process information.

Ecological groupings

One of the most widely used approaches to assemblage characterisation is to group on the basis of ecological compatibility. This can be done by grouping taxa regarded as characteristic of a particular habitat or those which occupy a similar niche (in the class 1 sense, Box 3.3). With land snails, the groups comprise species of similar ecology, but they are not necessarily groups reflective of specific environmental conditions. A group of molluscs from woodland can have the same percentage of woodland species as a group from cave rubble or scree, but the associated communities of the two environments are quite different (Evans and Jones 1973). Kenward (1978) suggests grouping Coleoptera according to where they live (e.g. aquatic spp.) or what they do (e.g. obligate phytophages).

Two points of principle require discussion. First, this is based on the assumption that the modern ecology of a species is a guide to its former ecology. When a beetle species, for example, is put into a particular ecological group, there is an assumption that the modern niche is fundamental and not a realised niche to which the species is constrained by competition or other pressures. It may be evident from the distinctive ecology of the species or from a specialised bodily adaptation that a wider or different niche is not accessible to it, in which case the assumption may be a fair one. Equally, it may be that the species composition of a particular ecological grouping is fairly constant across the range of archaeological assemblages and closely matches that seen today. In this case, too, it is fair to assume that the species represented in that group today are not occupying a narrow, realised niche. The assumption may not always hold, however, and the possibility that species may formerly have occupied different habitats or niches is one weakness of ecological groupings (Fig. 3.1). This is one of the problems of the use of histograms (e.g. Fig. 15.4). The ecological significance of the groups and individual species can vary through time (Thomas 1985), so that a group reflective of woodland at one level may be reflective of a more specialised rubble or synanthropic habitat at others. The second point is that some ecological groupings mix habitats and niches. To stay with beetles, there is an important difference between 'phytophagous' which describes their niche and 'strongly plant-associated' which describes their habitat. Phytophagous beetles will inevitably be strongly plant-associated, so niche and habitat are linked, but the strongly plant-associated group can also include predators of the phytophages.

Phytosociology

Phytosociology was developed in Europe to describe present vegetation, with emphasis on which species were typically found together in which habitats. Its founder was Josias Braun-Blanquet, and the 'classic' school of phytosociology takes its soubriquet from the two universities with which the early development was associated, the Zurich-Montpelier school (Ellenberg 1988). Phytosociology is based on the principle that vegetation types in terms of floristic (i.e. species) composition recur as communities which are essentially related to the physical environment, especially soil, hydrology and climate. The recurring communities are named according to the two predominant species. A sample area is chosen within which there is homogeneous plant cover and a list of all the species therein drawn up, with a semi-quantitative estimate of their relative frequency based on the proportion of the sample area covered. This list is a *relevée* or *Aufnahme*. By grouping *Aufnahmen* and examining the degree of similarity between them, a hierarchical classification of plant communities is developed. The smallest units are 'associations', which are grouped into 'alliances', then 'orders', and so on to 'formations', which are as broad as, for example, 'forest'. The system appears to offer a ready-made structure for archaeobotanical analysis.

Phytosociological methods have been used at Maiden Castle, Dorset (Sharples 1991) and Runnymede Bridge, Surrey (Needham 1991). Hall and Kenward (1990) used associations to group material from urban archaeological sites such as York, whilst Jones (1992) has shown that the recognition of particular associations in material from south-eastern Europe may be useful in understanding the cultivation of different soils and their weed floras. Greig (1988) applied phytosociology to the study of ancient haymeadows, and there is no doubt that the structure of associations and alliances is useful.

Latterly, however, a note of caution has been sounded, notably by Küster (1991) and Hillman (1991). As we have pointed out elsewhere (pp. 15–16), archaeology tends to sample the modified and operational environments within which human interference with the vegetation can almost be regarded as inevitable, and human transportation of potential fossils, such as seeds and pollen, must be seen as highly probable. Between the original plant community and the archaeological sample, then, is the filter of human behaviour. Furthermore, phytosociological attribution of the abundance status of different species within an association is based on coverage of the ground at the time of sampling, whilst archaeological data are typically numbers of macrofossils or pollen grains, which cannot be sensibly converted to area of ground covered. There is a problem of resolution with comparing modern data from a known area sampled on a particular day with ancient, time-averaged data obtained from an imperfectly understood catchment. It can also be argued that the whole Zurich-Montpelier procedure is fundamentally flawed because of the stress which it places on the continuity of association of species, with little regard for (a) the exigencies of colonisation or local extinction at any one sampling locus, (b) the possibility that the original *Aufnahme* may represent a community which has undergone recent disturbance and is thus at disequilibrium, or (c) the reasonable premise that species of plant are seldom mutally cooperative and thus that an *Aufnahme* as recorded represents only a temporary mutual accommodation between otherwise competing species. In many ways, the Zurich-Montpelier procedure is open to much the same criticisms as have been levelled at the theory of vegetational succession (Putman 1994, 94–100).

Mutual Climatic Range Method (MCRM)

If two or more species in a subfossil assemblage have present-day ranges which overlap, then the climate in the area of overlap (the Mutual Climatic Range – MCR) is a closer reflection of the ancient environment than in the area of any one of the species alone. For an assemblage in which species 1 to 3 are present, the area of overlap of the zones of tolerance for parameters 1 and 2 is the most likely range of values (Fig. 13.4). This principle is used especially in the climatic assessment of insect data. 'In general the MCR will be smaller and the palaeoclimate deduced more precisely, the larger the number of co-existing species' (Atkinson *et al.* 1987). The method uses temperature-sensitive coleopteran species as a guide to past temperatures. Species which show a climatic influence on their modern range may be responding to climatic attributes like summer maximum temperature, winter minimum temperature or annual temperature range, so that any one species is likely to be a problematic climate indicator. However, a suite of species occurring together may provide overlapping information which gives a good general guide to the thermal climate. MCRM seeks to quantify these overlaps, and concentrates on the presence or absence of species from a sample, thus avoiding problems

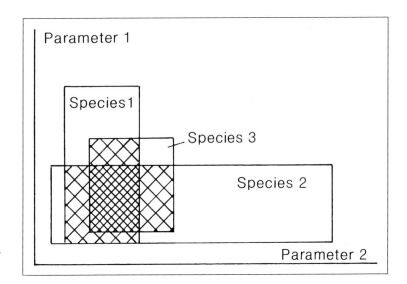

Figure 13.4 Principles of the MCRM.

of local abundance through factors other than climate. The modern distribution of species in the ancient sample can be determined with respect to two axes, typically the mean temperature of the warmest month, and the temperature range between the warmest and coldest months. By plotting the climatic tolerance of several species together, an area of overlap should be apparent (Fig. 13.4). Calibration using MCRM to 'reconstruct' modern climatic parameters for particular sampling sites indicates that the method is quite robust and gives reconstructed climates close to those obtained by direct record.

The method has been used to good effect to reconstruct temperature regimes, particularly for the later part of the Last Glaciation, the Late-glacial and the early Holocene (Coope and Brophy 1972; Coope 1977; Lowe and Walker 1997). The applicability of the method in mid- to late-Holocene settings, where human influences on biogeography might be expected, and synanthropic behaviour might have extended the range of some species, must be regarded as problematic.

It is sometimes said that some of these methods simply involve cleaning up the data prior to interpretation. But this is not so. These methods involve a degree of interpretation and it is important we realise this, otherwise important information may be missed about general or unanticipated aspects of the data in our rush to identify associations which we think are more interesting. Even so, spatial associations are not to be seen as causal without further analysis. With nearest neighbour analysis they can be related to the physical environment or to competition between settlements or populations; with small artefacts, distributions are often related to trade or exchange. In general (aside from uneven fieldwork or post-depositional factors), one grouping may affect another or *vice versa*, as with a plant distribution being caused by a particular soil type or the soil being affected by the vegetation. There may also be an unidentified step in the relationship, or both distributions may be related to a third factor, like aspect, and not themselves causally related at all (cf. Hodder and Orton 1979).

FOURTEEN

POSSIBILISTIC STUDIES

Possibilistic studies *suggest* interpretations of past environment and human activity. They use specific observations or general laws, not always from the geographical region of the archaeological study, to enhance the interpretation of archaeological data (Fig. 14.1). This is analogy of human rather than natural processes, although there is no sharp boundary between the two. It is the difference between monitoring the weathering of a ditch cut to simulate an earthwork and that of a natural cliff. There are four main areas of use.

1. They show a norm of behaviour based on maximisation of resources and technology. This can be compared with the actualistic data and the differences explored: 'comparison by anomaly' (Gould 1980).

2. They indicate specific possibilities, especially from present-day subsistence farmers and hunter-gatherers outside the experience of Western archaeologists: 'formal analogy'.

3. They indicate possibilities under experimental conditions of environment and technology, as with rates of forest clearance with stone axes or the effect on soil of ploughing.

4. They show rules of behaviour under certain conditions, although some of these are so general as to be not very helpful: 'relational analogy'.

There are two opposing scales of reliability, one in terms of the link between input and results, and the other of reality (Hodder 1982a, 30). Mathematical modelling and computer

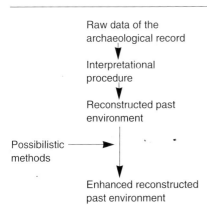

Figure 14.1 The position of possibilistic methods in interpretation.

simulation are most reliable in terms of input and results but distant from reality. After this, there is increasingly less control over the link between what is done and what is observed but an increase in the reality of situations: laboratory observations on scree formation or the movement of river gravels in flume tanks; outdoor experiments where there are so many factors that not all can be measured, such as the growth of blue-green algae on clasts in experimental earthworks; modelling vertebrate body decomposition (chapter 6); spatial modelling of settlement and resource use; experiments with grain storage and other aspects of food production; and formal analogues from modern communities where the symbolic use of land and material culture have no general basis.

HUMAN-LIFE STUDIES

Human-life studies are based on data that have not been collected specifically to assist in the interpretation of the past (cf. ethnoarchaeology, p. 185). They include ethnohistory, recent observations of communities whose life-style is now changed or lost (e.g. the Tasmanian Aborigines), and documentary records. A part of ethnohistory are the links between past and present where we believe that changes of environment and economics have been minimal – the basis for Gould's (1980) 'continuous analogy'. As models for the past, human-life studies tend to consider communities living at or beyond the edges of intensive, mechanised agriculture, industry and urban life (Lee and DeVore 1968; Orme 1981; Hodder 1982a). However, the whole spectrum of present-day or recent communities and environments must be studied, from complex states, to village and hunter-gatherer peoples, including those in middle latitudes and temperate climates (e.g. Halstead and O'Shea 1989).

A major caveat is that there are likely to have been basic changes in the human niche through the period of human development: 'Presently observable cultural systems do not necessarily reflect the total range of prehistoric ones' (Gifford 1980, 94). We need to be careful before we equate any human activity with environment and set up rules of behaviour from one or a few instances. Even then, general rules, such as the correlation of agriculture with particular climates, only tell us what humans could do, not what they did. Possibilistic methods often assume that interactions between humans and environment are predictable for a given technology, culture and environment (physical and social) (pp. 5–6; Steward 1955), and with maximisation of productivity and a functional relation with the environment as prerequisites of their use. Imbalance between expected (in terms of functionality and optimisation) and actual behaviour (what we find on and in the ground) can be explored and used as a basis for interpretation (cf. Gould 1980). A pattern of settlement territories based on geographical modelling of distance between settlements can be compared with boundaries known on the ground or from documentary data (chapter 9, p. 110). But people do not map themselves onto the physical environment, using all known and usable resources, even with the relevant technology and knowledge. There will be an interplay of functional factors like competition and differences in the timing of settlement establishment (which can affect the size and shape of territories), and also of more cryptic reasons such as social and religious taboos on certain foods. For example, fishing and the exploitation of lake resources generally in the *gal dies* (poor men) class of the Kenyan Dassenetch is considered demeaning by the pastoralist class and excludes the males from important ceremonies and social roles (Box 14.2; Gifford 1980, 96).

Equally, people need not maximise time and effort in collecting and producing food. Maximisation anyway is difficult to prove because there are benefits and costs which are hard to identify, even in living communities (Halstead and O'Shea 1989). Key benefits are relaxation and therapy, mediated through fire, animals, music and conversation. Burning of vegetation is a general feature of farmers and hunter-gatherers. Use of fire is claimed in a Neanderthal context in south-western France, and structured hearths, probably as social foci, were widespread by the end of the Last Glaciation. But no generalisations can be made as to the use of fire, other than that people were probably casual about putting it out (Gould 1980). Animals are a constant association of most human communities. Dogs are known from the Near East by *c.* 12,000 BP and from north-western Europe by *c.* 10,000 BP; they were brought to Australia at some time after 12,000 BP and to North America by *c.* 8000 BP (Davis 1987). Dog burials, with or without people, are common. In Australian aboriginal societies dogs are a hindrance in hunting, they have no use in defence, they are not used as food in any general or exceptional strategies, and they are often a nuisance around the campsites when they are mercilessly slain. Yet they are named and care is taken over their well-being, such as removing dangerous bones from their reach, although they are not directly fed (Gould 1980).

Conversation, ceremonies and rites are crucial not just as therapy but also for the perpetuation of knowledge and maintaining its accuracy, especially of resources and how to cope with risks like food shortage. These can take place within individual communities or at large, often annual, gatherings. Accrual and maintenance of information about resources through traditional forms of procurement and exchange also cut search costs and buffer hardship. They are embedded in all kinds of activities from noting particular plant foods when hunting, to widespread exchange networks covering tens of thousands of square kilometres. People spend a lot of time apparently, but only apparently, doing nothing.

There are also hidden costs, as where there is a substantial input of energy into the maintenance of strategies which are used only infrequently if at all (Halstead and O'Shea 1989), or those provoked by religion, ideology, financiers and banking in intensively industrialised and capitalist states (Newton and Porter 1988), or hierarchies in communist ones which drain the economy and benefit a bureaucratic elite.

Mobility in all societies beyond individual settlements makes it hard to identify territories, the land used, and thus whether maximisation was taking place. We can point to particular groupings of material culture or land-use, but these are usually only a part of the territory of a particular group or may cut across two or more territories (Hodder 1982b). Much of mobility is about spreading and minimising risk, either by having lands in several different areas, using different areas at different times, or by establishing contacts with other communities (social storage) (Halstead and O'Shea 1989). In particular, there is a range from mobility to sedentism in both hunter-gatherers and farmers. Some hunter-gatherers are more or less sedentary, as with some west coast North American and south-eastern Australian groups living in productive ecosystems. Others, like the Nunamiut Eskimo, move over catchments of several thousand square kilometres during a period which may occupy the entire lifespan of its members (Binford 1978).

Equally, while many farming communities are relatively sedentary, some, as with agro-pastoralists living in harsh environments, have complex systems of different residences, subsistence types and mobilities (Cameron and Tomka 1993). There is also mobility in rural communities and towns, with different kinds of regulation, from labourers, through land-owning peasants, higher echelons of local squires and landowners (gentry families, often acting as an unpaid civil service), to aristocracy and state. Priestly and aristocratic strata, with lands scattered over wide areas and variously managed, operate across these levels. Different land tenures (e.g. rented, owned, acquired *vs.* inherited), different qualities of control (e.g. fiscal, teneurial, seigneurial and state), lead to diversity of land and mobility, none of which may be related to the physical environment or to clear economic exigencies.

Underlying all of this, and part of the unpredictability in human land-use and economy, are history and tradition. Use of particular areas of land in particular ways relates to the past of the land – whose it was, how it was used, what state it is in – and to the past of the people currently using it. Social stigmas in the eating of certain foods may be similarly historically embedded. The influence of history can be from the recent past and relate to food shortage, the prediction of environmental change being especially relevant to catastrophes. Storage, the maintenance of social contacts, exchange of food and diversification are all relevant in buffering against famine or harmful conditions like flooding (Halstead and O'Shea 1989; Brown 1997). In the longer term, tradition is relevant for peoples who move from one region and environment to another and maintain their methods of land-use, psyche and regulatory mechanisms. Stone (1993) gives examples from eastern North America and the Jos plateau in Nigeria of the exploitation of land by sedentary intensifiers and mobile extensifiers contemporaneously in each case. The intensifiers manured and mulched, had extended field preparation including ploughing, planting on bush fallow and stump clearing, had high daily labour inputs and a lengthy farming season. In contrast, the extensifiers burnt and cultivated, but after a few years expanded onto new lands, with soil exhaustion being countered by moving the entire settlement; they also relied significantly on hunting and gathering. These different patterns, occurring in one place with the same climate and soil conditions, are embedded in the history of the groups concerned. There is no good analogue here for any model of spatial geography. Not only this, but their kinship patterns and regulation systems of dealing with land disputes were also tailored to the different agricultural strategies. Thus Stone (1993, 78) on the Nigerian example: 'Land disputes [among the Kofyar, the intensifiers] had traditionally been handled by village elders or clan members, procedures of no use when disputants were non-Kofyar.' And of the Tiv (the extensifiers): 'The segmentary lineage provided a ready principle for organizing pioneering units. . . . The lineage system was adept at mobilizing support for land disputes with non-Tiv and Tiv alike.'

Material culture is used to manipulate lives (Hodder 1982b). This may be overt, as in the exchange of raw materials and artefacts to establish social relations, in the positioning of houses in a settlement to indicate status, or in the establishment of physical boundaries around fields to indicate control. Conversely, it may be cryptic, like the structured depositions discussed by Hill (1995). Such practices can be embedded in the deep past and their significance unknown, even to the practitioners. Symbolling is also expressed when environments which have been unaltered physically by people become artificial by their use or adoption (Ingold 1986), or even just by thinking about them, and these 'affordances' can be brought into people's lives.

People make decisions in relation to the environment they perceive (Fig. 2.3), and we are beginning to build up some idea of this in terms of history, tradition, risk management and symbolism, much of which is hidden in life and in the archaeological record.

Ethnoarchaeology

Ethnoarchaeology is the study of the behaviour and material culture (including environments) of modern communities or their recently abandoned settlements with the specific aim of explaining the archaeological record. It is thus comparable to modelling taphonomy (Gifford 1980) (Boxes 6.1 and 6.2). Ethnoarchaeology also looks at landscapes and regions (Behrensmeyer and Boaz 1980; Gould 1980; Cameron and Tomka 1993), especially where change is taking place, as with the redirection of land-use or abandonment. There is a distinction between doing ethnoarchaeology in the specific areas of archaeological relevance – Gould's (1980) 'continuous analogy', Hodder's (1982a) 'direct historical approach' – and doing it anywhere and applying the results as general rules. These are exemplified in two case studies (Boxes 14.1 and 14.2), the former being used in the South Uist example, where the results were applied to archaeology in South Uist itself, the latter in the Dassenetch work where general rules were sought in order to understand hominid sites in the same area but from hundreds of thousands of years earlier and of another species.

Ethnoarchaeology is thus relevant to the following areas, amongst others:

1. The distribution of bones and artefacts at the intra-site to regional scale in relation to human and animal behaviour (cf. Box 6.2) (Behrensmeyer and Boaz 1980).

2. Preservation and burial, i.e. the formation of the archaeological record. Burial (and some destruction) by animal and human trampling and wind-blown deposits are common agencies, especially in semi-arid areas.

3. Abandonment at the site to regional scale (Cameron and Tomka 1993; Stahl 1995) (p. 76).

4. General aspects of human behaviour like the relation of territory sizes to language groups and distribution networks (McBryde 1986) and the interfacing between hunter-gatherers and their agro-pastoralist (e.g. Kent 1992) and colonial (e.g. Bird David 1992) neighbours.

Experiment

Experiment asks questions. It imposes conditions or contexts, against a background of specific needs as well as general theory, and it can be repeated (Coles 1973). Measurements of grain storage success, the efficiency of axes in tree felling, and other aspects of early technology, where the effects of one thing against another are tested, are

ENVIRONMENTAL ARCHAEOLOGY

Box 14.1 Howmore, South Uist: ethnoarchaeology case study 1

The study at the abandoned farm of Howmore, South Uist, Outer Hebrides, examined traces of particular activities in order to find signals for functional areas which could be matched with those in archaeological sites and thus reconstruct their functions (Smith 1996). The work was specifically directed at interpreting layers in archaeological excavations of 'occupation material' which could be variously occupation, primary or secondary midden refuse, field manuring, or specific activity areas and which are traditionally interpreted from a general analogue approach but without any modern parallels.

The farm had been 'abandoned' in 1938 (although the actual mode of abandonment was not specified, see below, Box 14.2, for the importance of this) and, although used since, the various contexts

Signatures of different functional areas and activities at Howmore, South Uist

Functional area	Activity	Signature
Byre	Accumulation of dung and bedding	High phosphorous and organic content
	Use of machair turves for bedding	Sand, molluscs and high pH
	Drainage	Drainage ditch
	Run-off dung and urine	High phosphorus
	Inhibition of organic decay	High organic, low pH and magnetic susceptibility
Barn	Storage of cereals	High density of phytoliths and a suite of cereal morphotypes
	Sweeping of floor	Absence of charred cereal grains
	Crop processing	Cobbled, clay-packed floor, inorganic, high magnetic susceptibility
Kiln	Processing of cereals	High density of phytoliths and a suite of cereal morphotypes
	Sweeping of floor	Absence of charred cereal grains
	Fires on the floor of the kiln	High magnetic susceptibility
Midden	Deposition of byre waste	Similar to byre deposits plus a distinctive suite of phytoliths
	Food waste removed by chickens and dogs	Chemical signals only
	Midden material put on fields and vegetable plot	Chemical signal only
Vegetable plot	Midden material used as fertiliser	Similar signal to midden
	Plant growth	Low phosphorus
	Cultivation	Homogenised soil and ploughmarks
Stack base	Storage of cereals	High density of phytoliths and a suite of cereal morphotypes
	Storage on stack base	Circular platform of large stones overlain by turf; sand (cf. the barn floor)

were untouched. Six were excavated and sampled; they were: the byre, barn, kiln, midden, vegetable plot and a stack base. 'Ethnographic' information about the building of the farm and details of the agricultural and domestic use were obtained from the original occupants. Analyses were for: phytoliths, snails, soils, macroscopic plant remains and architectural features of the structures themselves.

The results were interpreted in the ethnographic context of farming activities and the movement of resources within the agricultural system on traditional Hebridean farms. Disposal routes and post-depositional processes could thus be taken into account, and distinctive signatures for different functional areas established.

> **Box 14.2 The Dassenetch: ethnoarchaeology case study 2**
>
> This study was done to elucidate remains of early hominid activity in the area of east Turkana, Kenya (Gifford 1980). The Dassenetch people are agro-pastoralists living on the north-eastern shores of Lake Turkana. They have large home settlements which can be occupied for long periods of time, even permanently, and much smaller and more transiently occupied stock camps. There is also a *gal dies* (poor-men or fishermen) class (approximately 5 per cent of the total tribe) who live from fish and other resources of the lake, establishing simple camps at the lake edge.
>
> The ethnoarchaeological study was directed at three aspects: whole site preservation in relation to location and site type, artefact and food-animal bone distribution in relation to different types of site activity, and artefact and animal-bone survival in relation to the circumstances of deposition. In relation to site location, home settlements and stock camps are located along watercourses but high enough up to be out of reach of flash floods. Short-term sites, in contrast, especially those of the *gal dies*, are located in areas which are subject to frequent flooding. The more permanent sites, which reflect the main settlement type of the Dassenetch, are thus in locations which are less likely to be preserved archaeologically than the short-term camps of the *gal dies*, which are buried swiftly and with minimal disturbance, and there would thus be a bias in the archaeological record towards the settlements of a minority segment of the tribe. This is a general principle: sites exploiting special resources or used for specific activities – upland or floodplain grazing, mineral ore extraction or hunting – are often in areas which are prone to preservation, either, as in the Dassenetch case, because of burial, or because of the absence of later use and destruction (pp. 91–2). The second point about the Dassenetch sites concerns the distribution of artefacts and food-animal bone on the abandoned camps, which were planned during and some months after occupation. Some of the key results are as follows:
>
> 1. Social and political necessities of artefact curation lead to a low proportion of artefact to food-refuse discard; with regard to artefacts, the society is essentially a curatorial one (cf. chapter 9, Fig. 9.6).
>
> 2. Within the artefact category, most pieces were non-functional fragments.
>
> 3. Primary refuse was often moved into secondary refuse piles or middens away from the houses and hearths, and this is a feature particularly of the more permanently occupied settlements; *gal dies* sites were characterised by mostly primary refuse.
>
> Gifford also noted that some materials – primary refuse, and especially its smaller (<2.5 cm) elements – were buried in the surface soil by trampling (and probably wind-blown materials) during occupation. Larger fragments of surface bone were removed to middens or destroyed by scavengers and weathering, so that after twenty years the surface site inventory consisted of hearthstones, a few small artefacts that had been lost, and artefact fragments, but no bone. The subsurface levels consisted of numerous small bone and some artefact fragments, while middens contained larger bones that had not been scavenged.
>
> Various contemporary behavioural, syn-depositional and post-depositional factors thus result in a variety of spatial preservational patterns at the inter-site, site and micro-regional levels.

experiments (Robinson 1990). Opportunist observations of ditch silting are not experiment, but recording of silting in ditches built to specified dimensions, orientation and subsoil, where measurements are made at regular time intervals in recorded conditions of weather and biological activity, are (e.g. Bell *et al.* 1996). The start conditions of experiments are crucial. In monitoring the weathering of a freshly exposed surface as in an experimental earthwork ditch, we need to know not just the slope, orientation, type of rock and subsequent conditions of weather and vegetation growth, but also what happened to the rock before it was exposed. Rock surfaces have been lowered during the

Holocene by natural and human erosion, so that the frost-weathered layers present in temperate latitudes at the end of the Last Glaciation have been successively removed. This means that the surface layers in Neolithic pits and ditches were more friable and susceptible to weathering than those in experimental earthworks today.

Experimental farms at Lejre in Denmark and Butser (Hampshire) on the southern English chalk (Reynolds 1994) have studied land preparation, crop production and yields under different soil-preparation conditions, grain storage and food preparation. These experiments give data on the dynamics and bioenergetics of agriculture, especially the possibility of growing particular crops under particular conditions, potential yields and conditions of storage. Experiment is linked with evidence on the ground or from documents. Van Zeist *et al.* (1976) found certain crop seeds in Friesland *Terpen* excavations, and since the most extensive soils in the area were brackish and seemed unsuitable for crop growth, experiments were done to see if the types of crops found in the excavations would grow on them. The experiments were generally successful. Rates of growth were measured, yields monitored and the accompanying weed floras recorded. Young cattle were particularly destructive of the experimental crop in one year. Excavations revealed that fields in the area in which the crops might have been grown had enclosing ditches which could have excluded cattle and so protected the crops.

An important area of experiment is recording the effects of processes on soils and sediments, from the macro-evidence of erosion rates to the micromorphological evidence for different forms of land preparation (Courty *et al.* 1989; Macphail *et al.* 1990; Bell and Walker 1992, 190; Courty 1992), which can then be matched with features from excavations.

Ethnoarchaeology, Experiment and Taphonomy

Ethnoarchaeology, experiment and taphonomy focus on processes between life and preservation, with emphasis on what happens at and immediately after disturbance, as when a ditch is dug, trees are felled, soil is tilled or land abandoned. They get away from the aseptic categories of site and settlement types, species of plants and animals, vegetations and soil types, and processing sequences of crops (e.g. Hillman 1981) and animals (e.g. O'Connor 1993). Instead, they home in on the vagaries and contingencies of everyday life, abandonment, dereliction and decay. They are particularly concerned with very short-time events, with what is going on at the surface, in surface distributions, just below the surface, and at the full spatial range from intra-site features to regions.

Ethnoarchaeology and experiment can be done in the same area, monitoring the same agricultural processes, recording their effects on soils and vegetation, and comparing them with archaeological remains. At the small spatial scale, the taphonomy of biological materials is an obligatory part of experiment and ethnoarchaeology. More widely, it links sites and regions. Pollen and bone dispersal and deposition must be studied in the context of environmental change on the sites where some of the materials were recovered, and in the areas where they are deposited and where other changes occur. Preservation of abandoned sites, perhaps by sedimentation, can relate to abandonment of regions, leading, say, to erosion (pp. 130–1).

Ethnoarchaeology, experiment and taphonomy are most productive at the local and regional scale. It is difficult to see how cross-cultural rules can obtain, especially for complex farming societies and in different regimes of climate and biology. An obvious approach would be to study modern or recent farms from different areas in the same climatic zone, e.g. one each from the Shetland Islands, western Ireland and Brittany, and then to do the same thing in different climatic and cultural regions.

Site Exploitation Territorial Analysis

Site exploitation territorial analysis (SETA) is used to suggest site function, subsistence and other environment relationships through a consideration of the near resources around the site. ('Site catchment analysis' was the term originally used in this way by Vita-Finzi and Higgs (1970), but this now refers to the total area of exploitation as indicated by actualistic evidence (see chapter 5) (Bailey and Davidson 1983).) The site exploitation territory (SET) is the area around a site most intensively exploited, especially for cultivation or gathering plant foods, and that can be reached within a short time. It is defined on the basis of observations on modern communities in terms of distance or walking time from the site: for hunter-gatherers a radius of 10 km and 2 hrs, for farmers of 5 km and 1 hr. Shorter distances and times have been used, e.g. for agriculturalists, 2 km (Ellison and Harriss 1972) and 10 minutes (Mytum 1988) for the most heavily worked land.

Before doing a SETA we need to know that the sites were contemporary, although this is not always possible to ascertain initially, and anomalies can show up in the analysis itself where there is much overlap of SETs. Also, the archaeological period, technology and broad economy must be known since it is pointless to consider the potential of resources unless they were exploitable in the first place. Past physical environments must also be known, especially in semi-arid areas where river downcutting and alluviation erode and bury surfaces (Box 10.2). In Mediterranean countries there was serious silting of bays in the later Holocene, so that many sites which were once on the coast, like Troy, are now inland with much agricultural land in their catchment (Bintliff 1981). Present-day land-use is not always a good guide. Tractor-drawn ploughs and chemical fertilisers allow areas to be put to arable which were formerly used for pasture. Documentary sources as recent as the 1930s may give information about traditional land-use, crops and yields, even in regions with a recent history of intensive mechanised agriculture.

A major precept is that SETA is done on a series of adjacent sites so that *relationship* of territories can be seen. Where there are considerable overlaps of SETs, or gaps between them, time or distance functions may be modified accordingly, while unusually large, small or irregular SETs may be interpreted in terms of their relationship with others in a regional land-use system, perhaps as areas of tactical importance (Box 14.3).

Results should also be assessed, where possible, against input from other possibilistic studies like the use of Thiessen polygons (below), and of actualistic data, especially of documents, excavation and field archaeology. Often this shows that actual SETs are not equivalent to theoretical ones, and the reasons for this can be various. Settlements, although contemporary, can be of different ages of establishment (Ellison and Harriss

Box 14.3 SETA in the Upper Palaeolithic of northern Spain

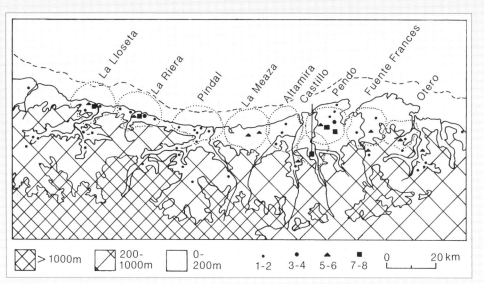

Topography and archaeological sites of Cantabria. Sites are weighted according to the number of periods represented. Dotted boundaries represent 2-hour SETs of major sites.

Bailey and Davidson (1983) did two regional SETAs of the Upper Palaeolithic in Spain, one in the area of the eastern Cantabrian mountains and adjacent coastal plain, the other in southern Valencia. All the sites were caves or rockshelters, and it was considered that the lack of open sites did not skew the data (p. 96). Sites used for SETA were 'site clusters', defined as sites so close to each other that they have virtually identical SETs. The SETs were defined by time to allow for the terrain; this was done mathematically with reference to contours, since walking the study area would have taken many months. The skewing effects of different types of vegetation and the distribution of snow and ice are hard to take into account, especially as transportation methods, e.g., snow-shoes, skis, sleds, are unknown.

SETA was used to examine sites' tactical use both individually and in relation to each other as a part of a regional economic strategy. The SETs of some sites encompassed wide areas of land and the sea (formerly land during the Upper Palaeolithic because of the lower sea-level), although these became constricted inland because of steep terrain. SETs further inland were more irregular and smaller. With reference to the ideal, circular SET, the actual ones (as calculated with reference to topographical variation) were expressed as follows:

Coastal plain (plus sea area)	60%
Inland lowland	37%
Upland	21%

It was also the case that the sites with the larger and more regular SETs were the larger ones – as defined by the numbers of archaeological periods (although artefact density and diversity and bone density might have been better measures of this). The coastal-plain sites can be considered as being regularly occupied, taking advantage of the better climate and the animals of the rich grazing lands, the inland lowland ones as sites exploiting particular resources at certain times of the year, e.g. animals of seasonal grazing areas, and the upland ones as tactically placed blocking or game intercept sites close to ravines specifically for hunting animals moving between the uplands and lowlands; and, indeed, it was noted that the fauna on these inland sites were dominated by ibex (*Capra ibex*).

> ### SETA in the Upper Palaeolithic of northern Spain
>
> The site clusters are spaced evenly in relation to each other. Eight major clusters have almost contiguous SETs, and this is not a result of the availability of suitable caves, because others were available with no evidence of occupation. This suggests that the SETs are real, that this area of the coastal zone was preferred, and that there was intensive exploitation of it.
>
> The site at Castillo was especially important since it is large (repeated occupation from the Middle Palaeolithic to the Neolithic) and its *catchment*, as indicated by marine shells, includes the Cantabrian coast (15–22 km away) and the Mediterranean (at least 470 km), although the Mediterranean shell is from a late (Azilian) level. Art is also included in the definition of the SC in terms of stylistic similarities, with contacts indicated with the coast to the north and the *meseta* to the south, some up to 500 km away. Castillo is additionally important in being an inland site with an irregular SET. It is well placed to control a major communication route between coastal plain and the inland *meseta*, and it was suggested that this site cluster was part of an integrated regional system with site clusters on the coastal plain for hunting.
>
> The site clusters and their SETs can thus be seen in relation to the exploitation of local resources, and their function interpreted accordingly. They can also be seen in relation to each other as being closely packed in an area of rich resources and intensive exploitation. And they can be seen as groups of site clusters working together in strategic systems of resource exploitation.

1972), with the later ones being more asymmetrical, smaller (or larger), and having poorer resources. Irregularities of resources can lead to irregularities of territories, providing strong clues to a particular resource exploitation. Change may also take place within established patterns with more powerful settlements expanding at the expense of weaker ones (Bintliff 1985). Exploitation of usable resources is rarely uniform, and this can lead to lopsided territories. For example, some agrarian coastal communities seem to make little use of the rich and accessible marine and littoral life of 50 per cent of the potential SET (e.g. Benson *et al.* 1991).

Lessons can also be learnt from observations of present-day communities, which seriously disrupt the expected pattern of decreasing intensity of resource use away from the settlement. Agropastoralists of Estancia Copacabana in south-west Bolivia have an aggregated main residence, three isolated main residences, several agricultural residences and numerous pastoral residences, each group at a successive distance from the main residence and the whole covering 1,000 sq km; additionally the pastoral residences may be seasonally or more episodically used (Tomka 1993). The Raramuri of Rejogochi, in the south-western Chihuahua, Mexico, have warm-weather residences where the household spends most of the year, agricultural residences which are similar but nearer the fields and occupied for a shorter time, and winter residences for the coldest three months of the year. This is complicated by the facts that some households are non-mobile, with year-round occupancy of some residences, and that there is movement of some households beyond the community to reside in others (Graham 1993). Agricultural residences in these two systems allow the more intensive exploitation of arable further from the main settlement than would otherwise be the case. For hunter-gatherers, there may also be intensive use of areas well beyond the SET by means of storage, trapping and mass slaughter, and sites

ENVIRONMENTAL ARCHAEOLOGY

Box 14.4 A GIS case study

The techniques of GIS were used to examine possible spatial correlations of sites, soils, and vegetation for a narrow period of time in part of the Pennine hills, UK (Bescoby 1997). Archaeological survey around the village of Malham revealed high densities of settlement sites and field boundaries apparently dating from the later Iron Age and Roman periods. The aim of the GIS study was to explore possible environmental settings for these sites, and thus to contribute to the interpretation of the spatial and functional relationships of the sites. The first step was to establish data layers which mapped the solid and drift geology, the topography, drainage and altitude. From these parameters, and using data from soil surveys in the research area, a map of postulated unmodified mid-Holocene soils was generated. Taking a regional pollen diagram to provide an inventory of taxa, and using known correlations between plant species associations and factors such as soil, slope and drainage, a provisional vegetation map was produced. This assumed that no clearance of woodland had taken place, and so the mapped vegetation was a theoretical mid-Holocene 'climax' vegetation for each of the different terrain units.

The landscape was then 'cleared' for agriculture, making simple assumptions such as that low gradient areas with deep, well-drained soil cover would have been cleared in preference to steep slopes or areas with thin lithosols. On the basis of modern observations within the area, the cleared areas were 'vegetated' with the most plausible grassland community, assuming that much of the clearance was to provide grazing for sheep and cattle, rather than for arable agriculture. This

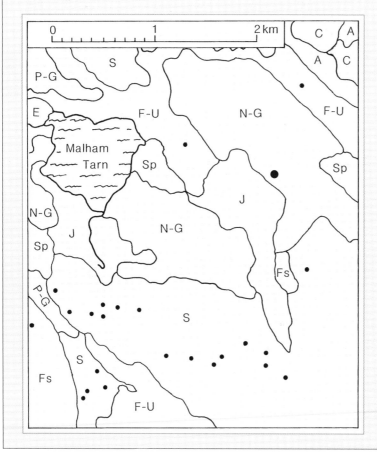

Postulated mid-Holocene vegetation, with location of Iron Age and Roman sites; the large dot is a Roman marching camp.
Grassland:
S = Sesleria,
N-G = Nardus-Galium,
Fs = Festuca.
Woodland:
F-U = Fraxinus-Ulmus,
A = Alnus.
Heath and mire:
C = Calluna *(dry)*,
E = Ericales *(wet)*,
Sp = Sphagnum,
J = Juncus-Molinia,
P-G = Pteridium-Galium.

A GIS case study

Romano-British structure, excavated in its surviving entirety.

assumption was made because much of the surviving archaeology appeared to be structures to facilitate the moving and penning of livestock, rather than the delineation of fields. This step in the analysis produced a number of large areas of grassland, some dominated by mat grass *Nardus stricta*, others by finer-leaved taxa such as *Festuca* and *Agrostis* species, and *Sesleria albicans*.

When the known later prehistoric and Romano-British archaeological sites were plotted onto this postulated vegetation map, the sites with clusters of 'roundhouses' and small enclosures showed a distinct association with the edges of the *Festuca/Agrostis* and *Sesleria* communities, and seemed to avoid the *Nardus* grasslands. One notable exception to this was a Roman marching camp, which may therefore have been deliberately sited on land of low utility.

The exercise 'proved' nothing, but it did show that the distribution of the sites was consistent with the developing model of a mainly pastoral landscape. Furthermore, discussion of the GIS exercise with farmers in the region added the observation that *Sesleria* and *Festuca/Agrostis* grasslands are particularly favoured for providing a good 'bite' for spring lambs. As the possibility of seasonal transhumance was also being considered, this was a valuable insight. In all, a fairly simple GIS exercise raised a number of intriguing possibilities which have helped to develop and refine the archaeological interpretation of the region. The process is a recursive one – our ideas about the archaeology drove a number of the decisions which were made in setting up the GIS base – but the close association of sites with a specific grazing land was unexpected, and was therefore a valuable contribution to the project.

may lack self-contained territories where they are part of a regional economic strategy, as for the killing or interception of game (Box 14.3).

THIESSEN POLYGONS

Thiessen polygons (Kopec 1963; several examples in Bintliff *et al.* 1988) are one of several spatial geographical methods which allow assessment of the apportionment of land between adjacent territories. They show theoretical boundaries based on equal sharing of land between neighbouring sites, with each site using only the land geometrically close to it. The sites must be in contemporary use, on the whole permanent, of similar kind, and the pattern complete. The Boeotia survey (Bintliff 1985) is a good example of the combined use of Thiessen polygons and several other sources of actualistic data. Known boundaries of city states from the fourth century BC (late Classical period) were compared with distributions of 'prime-quality light arable soil', Thiessen polygons and SETs. Divergences between the known boundaries and Thiessen polygons were primarily due to the known territorial expansion of the more powerful states (especially Thebes) onto land of weaker neighbours. Blank areas between the SETs and the Thiessen polygon or actual boundaries may have been occupied by towns for which there is as yet no field or documentary evidence (cf. Fig. 15.5). Bintliff further discusses the data with reference to documented population figures and land-use.

There are many other methods of spatial analysis which suggest the partitioning of land into exploitation territories, the ways these might have been utilised and their relation to each other. Amongst these are nearest neighbour analysis (chapter 13), central place theory (Hodder and Orton 1979) and game theory (Gould 1963). When used with actualistic data, these methods may show departures from a functionalist economic norm or from the maximisation of resources, for example in asymmetric territories and small territories below catchment thresholds (Bintliff *et al.* 1988, 93). These can be assessed in terms of missing sites, sequences of establishment, variations in function and the development of differential power. They help us to understand the contribution of different factors to distributions, especially those other than the physical environment.

MATHEMATICAL MODELLING AND COMPUTER SIMULATION

The advent of Geographical Information Systems (GIS) has allowed ever greater sophistication in spatial modelling. We can model single settlement phases under different parameters of, for example, crop yields, rainfall and technology, or model the direction and rate of changes that take place from one period to another. We can model global sea-level for a particular period of the past or future if we feed in information about the positions of the sun and moon (and planets if they are felt to be important), the global glacier and ice-sheet budget (i.e. the rate of melting and accumulation, themselves related to solar energy), the amount of carbon dioxide in the atmosphere, the strength of the Earth's magnetic field, and other factors that we think are relevant. Allowances can be made for different factors and quantities, and different models or end results can be produced. Nothing is definite, but this kind of modelling serves a valuable heuristic role in suggesting possibilities, and perhaps in showing the power of different forcing variables under different combinations of parameters. We can also model the rates of spread of particular species and communities as environments change. We can see patterns of infilling and circumstances where competition may have been significant. Again this is all possibilistic and depends on the user-defined start conditions.

A slightly different example is given in Box 14.4, where a detailed terrain model of a region was used to investigate the spatial correlation of archaeological sites and postulated plant associations.

Fifteen

Case Studies

In this chapter, we review examples of the archaeological investigation of the human environment. The examples illustrate a range of spatial and temporal scales, different research priorities and a range of sources. Two of the examples are ones in which we have had some first-hand involvement, and for which we therefore take a share of any criticism.

Northern England: The Roman Impact

The region of northern England and the period of the Roman occupation have been chosen as an example of the difficulties and pitfalls of trying to study the 'impact on the environment' of a particular series of historical events. Much of what we set out to do is to examine the mutual influence of people and environment, and it would seem that a clear, historically recorded phenomenon such as the Roman occupation of Britain would serve as a fruitful area for research. So it does, but principally by raising difficulties and ambiguities which give us much cause for thought when considering the 'environmental impact' of peoples who lack a detailed historical record.

Northern England (Fig. 15.1) is topographically diverse, with a major area of high ground based on the north–south trending anticline of the Pennines. These hills are bounded by the low-lying Vale of Mowbray and Vale of York to the east, and are deeply cut in the west by the Vale of Eden. The history of the Roman occupation of this region begins with the founding of forts at Malton and at York in AD 71, followed within a few years by the founding of what was to become a major centre at Carlisle. Around AD 125, the northern boundary of 'Roman' Britain was delimited by the construction of Hadrian's Wall. Despite the subsequent building of the Antonine Wall (c. AD 145) further north, Hadrian's Wall became the effective boundary between Roman and native-controlled regions. The region thus holds two of the most important Roman centres of settlement in Britain – York and Carlisle – and a major defensive boundary. It is not surprising, therefore, that much of the archaeological investigation of the Roman period in northern England has focused on these three locations.

We cannot assess the impact of the Roman occupation without a fairly clear view of the environment into which the Roman armies moved. There are also questions of scale and rate of impact: what was immediate, and what was the consequence of subsequent consolidation of settlement? How far did events around the major centres such as York and Carlisle affect the environment further afield? The study of this topic thus reduces to a series of questions about the scale of our study and about the resolution of our evidence when compared with the high chronological resolution provided by historical dates.

CASE STUDIES

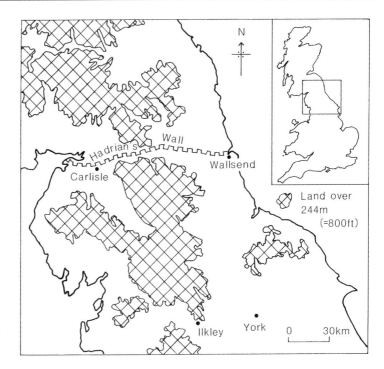

Figure 15.1 Northern England, with sites mentioned in the text.

The issue of dating has been particularly crucial in discussions concerning the construction of Hadrian's Wall and the forts associated with it. Although much of the surviving monument is built in stone, it is clear that very large amounts of timber were utilised in the original construction and in the construction of early phases of buildings which were subsequently replaced in stone. In addition, large amounts of wood were used in the manufacture of artefacts and in the provision of firewood and charcoal. It seems an obvious postulate, therefore, that the construction and garrisoning of Hadrian's Wall would have led to substantial clearance of woodland in the region. Note that two assumptions are being made here. The first is that structural timber would have been obtained locally, rather than imported to the region, and given the volume of timber required, that seems a fair assumption. The second assumption follows from this, namely that there was timber available in the region, that substantial clearance had not already taken place.

Dumayne and Barber (1994) set out to investigate this issue through the examination of pollen profiles from three mires in the vicinity of the Wall. At two of the sites (Walton Moss and Glasson Moss), there is a fall in the relative abundance of arboreal pollen beginning at a level in the profile which, from interpolation between radiocarbon dates, can be dated to the pre-Roman Iron Age. Taking the relative fall in arboreal pollen to indicate clearance of woodland, this appears to have begun in the Iron Age and to have continued through the Roman period. The third site, Fozy Moss, shows 'little human impact', until there is a quite abrupt rise in grass and Cyperaceae pollen, with a corresponding fall in arboreal, largely alder and birch, pollen. Values for the weed species *Plantago lanceolata* increase around the same point in the profile. There are only three radiocarbon dates published for Fozy Moss, and the one closest to this event in the pollen profile gives a date of 1820±45 bp. The 'most probable' calibration of this date is AD 130–245, which nicely fits the premise that the construction and garrisoning of the Wall from AD 125 would have led to substantial clearance. Less helpfully, the 95 per cent confidence range on this same date encompasses virtually the entire Roman period and a good deal of the preceding Iron Age.

Taken at face value, the Fozy Moss results seem to show pretty much the drastic clearance of a previously wooded landscape which might have been expected. We must note, however, that the radiocarbon determination allows a small but not insignificant probability that the events inferred from the pollen diagram actually pre-date the construction of the Wall. In terms of people and land, there would be a very big difference between major clearance in, say, AD 100 by local 'natives', creating a cleared landscape into which the Roman army moved nearly a generation later, and Roman clearance of a previously wooded landscape from AD 125. Unfortunately, the resolution of radiocarbon dates, and the interpolation of dates through pollen profiles, is insufficiently precise or accurate to distinguish between those two possibilities. If our concern was only with vegetation change, then those 25 years would hardly matter, but the precision becomes essential if our concern is to examine which people were having what environmental impact, and why.

At Walton Moss, towards the western end of the Wall, the pollen profile shows small, rapid fluctuations in the relative amounts of arboreal and herb pollen, which could represent brief clearance and regeneration events. Calibrated radiocarbon dates place these events between 90 BC and AD 215, again overlapping the late Iron Age and the Roman arrival. A similar profile from Fellend Moss, Lowtown, appears to show clearance calibrated to AD 2±45 years. Elsewhere in the region, and particularly to the east of the Pennines, inferred or radiometric dates for apparent clearance events range quite widely (Huntley and Stallibrass 1995, 48–65). Some are clearly pre-Roman: Valley Bog, Durham, has a clearance event firmly dendro-dated to 265±5 BC (Chambers 1978), and a number of Roman sites in the region show evidence of ploughing in soils underlying the earliest Roman deposits and structures. A particularly good example is at Wallsend, at the eastern end of Hadrian's Wall. Elsewhere, the first major clearance events are surprisingly late. The pollen profile from Moss Mire shows clearance underway by about AD 400±90, then continuing to around AD 520±90.

In the end, the question of Roman or pre-Roman clearance remains open, despite a considerable weight of data and a certain amount of firm assertion. Dumayne and Barber (1994) make it quite clear that they see the pre-Roman north as relatively sparsely populated, citing the paucity of evidence for prehistoric settlement in the vicinity of Hadrian's Wall. That starting assumption allows them to be more confident, one suspects, in attributing to the early Roman period clearance events which can be radiocarbon dated to somewhere from the late Iron Age onwards. McCarthy (1995) takes a somewhat different view. He questions whether the absence of evidence is really evidence of absence, suggesting that a predominance of pastoral, rather than arable, farming in the Iron Age may have rendered sites less visible, and pointing out that the lack of modern development in the region has limited the amount of planned excavation or casual discovery. McCarthy also offers a different point of view, based on the numerous recoveries made in Carlisle of waterlogged Roman timbers. The dendrochronological ages of these timbers seem to indicate that little oak timber older than about 250 years was used in the construction of Roman Carlisle. One possible interpretation of this observation is that there had been substantial clearance in the region during the Iron Age, so that few really old trees were available by the first century AD. There is a counter-argument: Dumayne-Peaty and Barber (1997) suggest that older oaks would have been senescent and unsuitable for construction timber, and so may have been used largely for firewood and charcoal.

A bulky material such as building timber we might expect to have been collected as close as possible to the point of use, but this clearly does not apply to many of the other materials which were extracted and utilised in Roman Britain. One of the other intriguing problems which this period presents is that of determining the degree and direction of spatial translocation of evidence. At one level, this is an economic question, a matter of establishing provenance and trade routes. In another way, though, there are questions which closely resemble those which we address when examining colluvial sediments. Just as colluviation may translocate sedimentological evidence of past environmental conditions a considerable distance downslope of the location where the particular environmental conditions prevailed, so the processes of trade and construction associated with the Romanisation of northern England brought about an increased degree of translocation of evidence of all kinds (cf. p. 69). We will illustrate and explore this point with reference to the record from York.

The first major environmental archaeology study undertaken in York concerned sediments within a Roman sewer beneath Church Street (Buckland 1976). The survival of fine-grained, waterlogged sediments offered the opportunity to study a very wide range of different fossil biota. Given that the assemblages were obtained from the fills of a sewer, with an uncertain history of water-flow and cleaning-out, and which may or may not have been fed by an aqueduct, which may or may not have originated outside the city, questions of origin and deposition were uppermost in this study. An obvious spatial dislocation had occurred between the living communities and death assemblages (pp. 68–9). The biota included plant macro- and microfossils, insects, ostracods, molluscs, small vertebrates and even sponge spicules. These different groups are more or less multilocational during life, and more or less likely to be transported by wind or water after death, before entering the sewer system and whatever further transport went on within the sewer. Live arthropods were noted in the sewer during its somewhat disagreeable excavation. The upper few centimetres of sediment were rejected for sampling on the grounds that they showed signs of pedogenesis, and biopedoturbation could have incorporated the remains of recently-dead arthropods and gastropods.

Modest amounts of pollen were obtained from three samples, and arboreal pollen predominated over herbs in all three. Eight different tree taxa were represented in all, though only alder, hazel and oak were abundant in all three samples. Taken at face value, the three samples appear to represent a well-wooded environment, with clearances dominated by grasses and low shrubs. Obviously this does not apply to the interior of the sewer, and it is hard to equate with the centre of Roman York, but we have little hard evidence on which to suggest whether the data show York to have occupied a position in a landscape still with substantial woodland cover, or whether the predominance of arboreal pollen is actually good evidence that the water supply for the sewer originated a considerable distance from the city. There is a large cultural component to the translocation of data: at least with geomorphological processes one can apply some knowledge of modern systems to assume, for example, that sediments do not move upslope unless they are wind-blown or driven onto sand banks and storm beaches. In the case of the Church Street sewer, there are simply too many possible explanations. As Buckland notes, the late Professor Fred Shotton even suggested that some of the arboreal pollen may have entered the sewer in human faeces, having been consumed in honey. The translocation thus involves bees and people even before the hydrology of the sewer becomes involved (Fig. 15.2).

Subsequent work in Roman York has raised two other examples of spatial translocation of evidence. Numerous finds have been made of fragments of peat, sometimes in a partially-charred state, and it is clear that this material was widely used as a fuel (Hall and Kenward 1990). Peat has also been recovered in appreciable amounts from Roman deposits in Carlisle. This adds one more complicating factor to the discussions surrounding the use of timber and wood in that city: if peat was used as a fuel, does that significantly reduce the postulated demand for firewood? By treating the peat fragments as samples in their own right, it has been possible to investigate the environment from which the peat derived: in the case of the York material, a raised mire with typical bog plants such as *Erica tetralix* and *Sphagnum imbricatum* is indicated. What the peat fragments cannot show is where this raised mire was located. Determining that important point becomes a matter for topographical analysis and a degree of intuition. In a way, the peat fragments are analogous to the fragments of translocated soils which are sometimes located through micromorphology analysis of buried soils in downslope locations, such as that reported by Macphail (1992) from Ashcombe Bottom, Sussex. In both cases, the translocated fragments give information about an environment other than that in which they have been deposited, but raise complex questions about depositional processes.

The other example of translocated evidence from Roman York concerns the occasional finds of macrofossils of saltmarsh plants. Given the riverine location of York, it is quite plausible that saline influence extended much further upstream than is the case today, with the development of saltmarsh areas quite close to the city. If that were the case, however, we might expect the insect and mollusc assemblages from the city to include appreciable numbers of those taxa which are typical of such habitats. Such taxa are not apparent, and

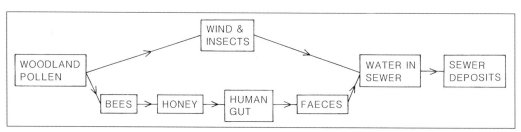

Figure 15.2 Two alternative routes and mechanisms in the deposition of arboreal pollen into deposits of the Roman sewer, Church Street, York.

the evidence for the presence of saltmarsh conditions somewhere within the catchment area of the city's archaeological deposits rests entirely on plant macrofossils. This implies the intervention of some transport mechanism which has acted to move plant remains but not those of invertebrate animals, and further implies that the saltmarsh areas were sufficiently far from the city to place it beyond the dispersal range of saltmarsh insects, in particular. The current favoured explanation is that saltmarsh areas some distance downstream of York were utilised for grazing land, with saltmarsh plants being redeposited in the city in dung or in the gut contents of slaughtered animals. There is other evidence for the stabling of horses in the Roman city: given the relatively inefficient digestive system of horses, it is quite conceivable that some plant macrofossils could survive transit through the equine gut whilst remaining identifiable. This is a slightly different version of Shotton's suggested means of transporting arboreal pollen into the city.

To sum up, what we know of the Roman environment in northern England depends very much upon our preconceptions about the pre-existing environment and social milieu. If we start with the premise that the late Iron Age north was sparsely populated and largely untamed wilderness, then it becomes easy to equate clearance events with Roman 'impact', or even to believe in woodland right up to the walls of Roman York. However, if we argue to the contrary, and believe that the archaeology of the non-Romanised peoples is just less visible, then much of the same evidence that would appear to support a substantial Roman impact can be used to argue something quite different. Perhaps, in the end, we can use the matter of perceptions to our advantage, by examining how the urban occupants and military garrisons appear to have perceived the land around them. For example, a number of the military sites in the area, notably the fort at Ilkley, show clear evidence of re-fortification around AD 200. That suggests a perceived threat, presumably from people, yet these people have left almost no archaeological record. That virtual invisibility at a time when northern England clearly had enough unfriendly natives to justify a re-fortification programme strongly supports the argument that a lack of later Iron Age sites does not indicate a lack of people, and therefore that some of the ambiguously dated clearance events could just as well have been underway before the legions arrived. In the end, researching the environmental impact of the Romans in northern England necessitates a move beyond pollen diagrams to questions of how people may have responded to particular perceived circumstances, and to trying to understand the social and environmental processes which may have acted to bring about the translocation and modification of different forms of evidence.

THE EARLIER NEOLITHIC OF THE ENGLISH CHALK

The earlier Neolithic environment of an area of central southern England (Fig. 15.3) has been investigated by molluscan analysis. The area is at the edge of the north Wiltshire chalklands. Apart from a seasonally intermittent stream it is largely dry but it is surrounded by more watered environments. There has been archaeological interest in the area for centuries because of a concentration of monuments (e.g. Stukeley 1743) and environmental interest because of the good preservation of molluscs (e.g. Evans *et al.* 1993). If we were considering just this area, these would be bad foundations for research

ENVIRONMENTAL ARCHAEOLOGY

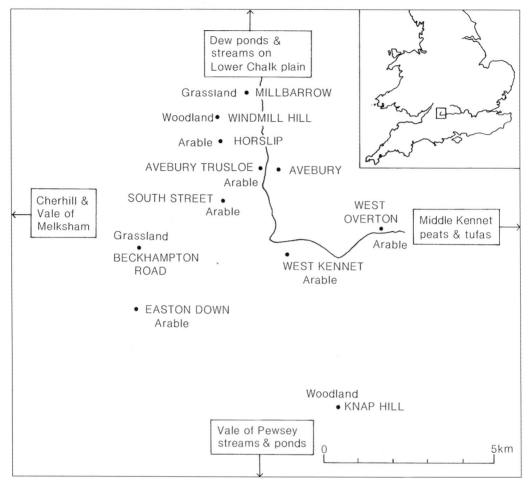

Figure 15.3 The headwaters of the River Kennet, central southern England, showing sites mentioned in the text and their main land-use; some of the arable sites were converted to grassland for a short period immediately before burial.

(chapter 8). But the investigation is larger, with work away from the monuments on the valley sediments, fieldwalking and excavation of adjacent sites off the chalk, and comparisons with other types of region – non-monumental and non-chalk – and Neolithic strategies (Whittle 1993) (cf. Fig. 8.5). The study shows especially how traditionally site-specific indicators can be used to build up a regional picture.

Procedural stages of molluscan interpretation

Interpretation of molluscan assemblages is exemplified using the soil under the bank of the Avebury henge (Fig. 15.4) (Evans 1972; Evans et al. 1985). The first stage is to assess the non-molluscan data, namely the general topography and the soil, sediment and archaeological contexts. The area is a chalkland plain above the level and influence of the

CASE STUDIES

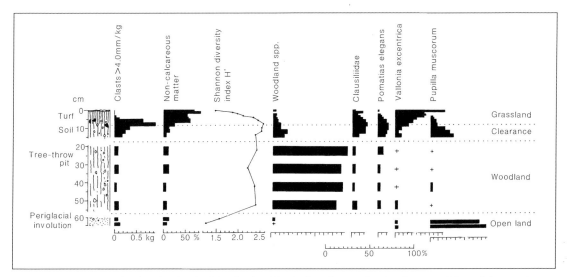

Figure 15.4 Buried soil under the Neolithic bank of the Avebury henge, showing features of the pedology and selected molluscan sequences (based on Evans et al. 1985).

ambient river flood levels and water-table, suggesting dry land. This is also indicated by the absence of a river channel or lake basin and the presence of a rendsina soil. The soil shows a sequence of dryland features – periglacial involutions, tree-throw hollows, ploughmarks and turf-line. Towards the surface there is a marked decrease in calcareous material, indicating decalcification towards the end of its life. It is buried under a substantial archaeological monument which itself suggests dryland and open-country. Establishment of the environmental origin of these features itself demands interpretation – through modern analogy, comparison with similar features under Neolithic monuments, and soil and sediment analysis. It also depends on molluscan analysis, so from the start we are not working independently of the molluscs.

The next stage is to consider the general nature of the molluscan faunas. Taxonomically, there are no freshwater species in any of the families represented. Ecologically, analogy with modern species also indicates dryland; a group distinctive of wetland habitats is absent, although some of the species would not be out of place in such habitats. Then we consider some general properties of the assemblages against the range of these in all dryland molluscan assemblages from Holocene chalkland environments in southern Britain. We could, for example, look at age distributions, although this was not done in this example because of the difficulties of separating juveniles from those of broken adults. Inventory diversity and the Shannon–Wiener index (H') (pp. 172–4) are more fruitful (Fig. 15.4), the latter showing practically the full known range, from below 1.0 to less than 2.8 (e.g. Evans 1991). There is very low diversity in the periglacial involutions, high diversity in the tree-throw pit, even higher values at the level of the ploughmarks, and then rapidly declining values in the soil itself. Now we can be more specific and suggest that if this range reflects the full range of chalkland environments and if inventory diversity is reflective of architectural diversity, ecosystem complexity and productivity (pp. 28–9) then we can infer something from the value of these parameters through the sequence. We can also describe the likely nature of the habitats in terms of vegetation and soil/surface disturbance, using the range in modern chalklands and fossil sequences of periglacial and Holocene environments as a guide.

Implicit in much interpretation at this stage is comparison of the assemblages through time; we may not be able to say what the environment was like at any particular level but we can say something about the direction in which it was going. Again this can be made more precise if we consider the range of possibilities. On this basis, the very low diversity assemblages from periglacial involutions suggest a treeless vegetation, the high diversity from the tree-throw hollow woodland, and the decreasing diversity in the main body of the soil profile to low levels at the surface increasing openness and impoverishment. Note that we are not talking in terms of processes or environments that are too distal to the data (pp. 169–70) like 'woodland clearance by man', 'grazing', 'abandonment' or 'climatic deterioration' at this stage. Can we be more detailed about particular environments? Here we look at ecologies of individual species and bring in specific information from outside the sequence from both modern and ancient assemblages. In the tree-throw hollow, some species of snail, in this instance *Vertigo pusilla*, *V. alpestris* and *Spermodea lamellata*, are fairly narrow in their ecology and suggest old, undisturbed woodland. There is also a virtual absence of open-country stenotopes, which suggests a lack of disturbance. In the open country towards the top of the soil there is a relative increase of *Vallonia excentrica*, a species which tends to favour poorer grasslands, so equating with the soil evidence of decalcification. What we should have done, which has never been done for the Neolithic grassland faunas of the area, is an ordination or similarity analysis so that we could feed back any groupings revealed by this into the raw sequences (cf. Fig. 13.3). The duration of the sequence is indicated by *Pomatias elegans* and *Discus rotundatus* which are present from the start (excluding the periglacial involutions) and which are not known in the British Isles before around 6500 BC (Kerney *et al.* 1980). At the recent end, dates for the construction of Avebury are around 3000 BC. Before leaving the profile itself, could any of the associations be taphonomic? Differential preservation of shells of *Pomatias elegans* and *Clausilia bidentata* in the main part of the soil profile is shown by the fact that they are all very worn and small apices, in contrast to those in the tree-throw hollow. Mixing of separate woodland and open-country assemblages at the base of the profile may have been caused by ploughing. And at the same level, the *Pupilla* shells probably derive from the Late-glacial involutions since many of them are the large cold-climate ecophenotype. So the shells at this level are of various origins and this is reflected in the very high diversity.

When we look at other soils on chalklands of the same general age and context we find similar sequences (Evans 1972; Evans and Simpson 1991; Whittle *et al.* 1993), indicating that the Avebury sequence is part of a more general pattern. A difference at Avebury is that the turf-line at the top of the soil shows clear signs of decalcification and age whereas the turf-lines in most other sites are younger and more calcareous. This is likely to be a function of the fact that the burial of the soil was much later at Avebury and is a point we return to below.

What the sequence means in terms of environmental processes at the site scale can only be assessed with reference to data from the area as a whole, to which we now turn.

The environmental sequence

References to sites are in Evans (1972) and Whittle *et al.* (1993) unless stated otherwise: contexts are valley sediments and buried soils and ditches of monuments. The pre-Neolithic vegetation was woodland, from the highest plateaux as at Easton Down to the valley bottoms as at Cherhill (Evans and Smith 1983) and in the Winterbourne and Kennet valleys. The high diversity of the fauna and the paucity or absence of open-country stenotopes is convincing for a completely shaded vegetation. The richness of the faunas of the lower infillings of monument ditches implies woodland close by, even though any tree and shrub growth in the ditches is a localised reaction to the specific conditions of the ditches and is secondary (Fig. 12.2).

The subsoil was not chalk rock but drift, often with a substantial non-calcareous element, as from loess or clay, and often infilling involutions. The natural soils were often brownearths or calcareous brownearths with B or Bt horizons, as shown by small fragments of residual material in tree-throw hollows (e.g. Macphail 1991). Outside our area, sequences of eroded soils in chalkland valleys often begin with non-calcareous materials, sometimes derived from loess, before becoming more generally chalky, a sequence which reflects in reverse the original soil profiles (Bell 1992). So several lines of evidence indicate the former presence of woodland and brownearth soils – the tree-hollows themselves, the snails and residual soil crumbs in them, snails in the ditches, and sequences of eroded soils.

The change to open-country mollusc assemblages and rendsina soils took place before the construction of the monuments. How was it caused, when did it take place? There is no evidence as to what was going on in the last centuries before Neolithic settlement although there are late Mesolithic artefacts from the area and there may have been woodland use in the early stages of the Neolithic. The change to open-country was probably caused by people and probably by Neolithic people although Mesolithic activity cannot be discounted (cf. Bush 1988b, for the Yorkshire Wolds); the critical horizons are at the bases of soils but the sequences are always highly conflated. The brownearth soils were probably converted into rendsinas during the Neolithic period by tillage as is seen especially at the South Street long barrow. Mixing of B and A horizons with highly calcareous material from the subsoil inhibited the formation of a B horizon (Duchaufour 1982). It also caused truncation of periglacial features and tree-hollows, seen in field sections and in abrupt changes in the snail assemblages, as at Avebury (Fig. 15.4).

Can we say anything about the spatial distribution and intensity of clearance and its duration (Fig. 15.3)? If we look at the buried soils underneath the causewayed camps (ceremonial monuments and feasting places) we find the faunas are woodland ones, although with some indications of disturbance, suggesting that the sites were cleared of trees more or less immediately before the monuments were constructed. At Windmill Hill (Fishpool 1992), the very stony soil suggests some disturbance, maybe by ploughing, but this was short-lived. The snail faunas in the ditches were woodland ones from almost the ditch bottom, suggesting that the woodland was close by and had not been cleared from a wide area. The sites were in areas away from settlements, and the woodland setting may be no more significant than this; or they may have been deliberately constructed in woodland as a kind of screen from the outside world. In other chalkland and limestone areas causewayed camps were also built in woodland as at Maiden Castle in Dorset (Sharples 1991), sites on the South Downs (Thomas 1982), and Crofton in the Vale of Pewsey, Wiltshire (Lobb 1995). Another sort of site, the long barrows, were built in open land, often arable and usually, where the evidence is detailed enough to show it, on the edge of it. Several sites cover the boundary between two different types of soils, one calcareous and disturbed, the other less calcareous or completely decalcified. At South Street there was a line of stakes, and although these had been removed before cultivation began they indicate the position of a boundary. At Easton Down (Whittle *et al.* 1993), there was a boundary in the snail fauna. Studies of modern snail faunas across vegetation boundaries show that they reflect quite faithfully vegetational differences (Fishpool 1992; Gardiner 1991) so the Easton Down evidence is interpreted accordingly as a division between scrub and grassland. In other areas, long barrows were also sited on the edge of arable, as in the Lincolnshire Wolds at Skendleby (Evans and Simpson 1991) and in Oxfordshire at Ascott-under-Wychwood.

Three features enable more detail to be added about the extent of woodland and the size and duration of the open areas. They are (Table 15.1) the vegetation and land-use of the area in which the sites were built, the rate of woodland regeneration measured by the first appearance of woodland snails in the ditch fill, and the duration of this secondary woodland (although the last is partly related to the timing of subsequent clearance). From these we can say:

Table 15.1. *Early environments associated with sites in the chalkland study area*

Site	Topography	Primary vegetation/ land-use	Rate of woodland regeneration	Duration of secondary woodland
Knap Hill (cc)	High down	Woodland	Fast	Long
Windmill Hill (cc)	Down	Disturbed woodland	Fast	Long
Easton Down (lb)	High down	Arable to grassland	Fast	Long
South Street (lb)	Down	Arable to grassland	Moderate	Moderate
Horslip (lb)	Down	Arable to grassland	–	–
West Kennet (lb)	Down	Arable to grassland	–	–
Beckhampton Road (lb)	Valley/Down	Grassland	–	–
Millbarrow (lb)	Lower Chalk vale	Grassland	Slow	Short
Avebury Trusloe (v)	River valley	Grassland/Arable	None	None
West Overton (v)	River valley	Grassland/Arable	None	None

cc = causewayed camp; lb = long barrow; v = valley bottom; – = no information

1. The causewayed camps were in the most continuously wooded setting and this is related to their primary and continuing function and perhaps to the deliberate maintenance of woodland.

2. The Easton Down long barrow was in the next most wooded setting, even though the area had been cultivated. This was perhaps a site peripheral to the main areas of cultivation and settlement as suggested by its high altitude and distance from water.

3. South Street (and perhaps Horslip and the West Kennet long barrow) saw some regeneration of woodland but it was not fully closed and did not last long. These sites are much closer to water and less exposed than Easton Down, and were probably closer to important nuclei of cultivation and settlement.

4. Millbarrow (Whittle 1994) and probably Beckhampton Road were on clayey Lower Chalk and clayey drift respectively, and the cold and heavy soils were probably less suitable for cultivation. They supported grassland (blocky structure, micromorphology, masses of fine rootholes, snail faunas), and the weak suggestion of not more than scrub regeneration, lasting only for a short time, reinforces this. These may have been in areas of permanent pasture, although Beckhampton Road is several kilometres from water.

In the valleys, at Avebury Trusloe and West Overton, a prehistoric dryland soil (Evans *et al.* 1993), uninfluenced (at least initially) by waterlogging, flooding or alluviation, extends across the valley, broken only by the river channels. This contains Mesolithic and Neolithic artefacts and is buried by alluvium which began to form in the late Neolithic. Micromorphology and the snails show woodland, woodland clearance and grassland and/or cultivation, with slight soil erosion just at the very edges of the valley in the form of micro-lynchets. There was, however, no woodland regeneration, perhaps because the soils were getting wetter or the valleys were used for intensive grazing, or both.

What else should we be doing?

Key questions about earlier Neolithic people in the area are what was their relationship with Mesolithic communities, what were they doing in woodland, where and how were they clearing woodland, and what were their subsistence strategies. We also need to ask where this evidence is going to come from, and at the moment only this question can be answered; namely by careful and informed excavation and analysis of valley edges where the valley sides abut the valley bottom, especially where slopes are steep enough to provide stratigraphy (through slight erosion) but not too steep that they could not have been cultivated (Evans *et al.* 1993). We also need to be examining dry valleys, first by coring to locate good areas of stratified Holocene soils and then by excavation. No such work has been done in the study area. Further approaches are the examination of areas where there are other concentrations of Neolithic archaeology on chalkland and limestone, such as the Lincolnshire Wolds and the Corallian hills of north-east Yorkshire, areas where other techniques like pollen and lake sediment analysis are applicable, and areas of continental Europe where earliest Neolithic strategies may have been quite different from those in the British Isles – in other words, the application of different scales and techniques, and in different environmental and cultural settings.

The absence of Neolithic erosion

A striking feature of the earlier Neolithic on the chalklands (and indeed in other areas of the British Isles) is that there is no evidence of erosion of cultivated soils, such as deep deposits of colluvium, except on a small and localised scale (Allen 1992). Clearance of vegetation and cultivation of the soil were not affecting the soil crumb structure, nor the level of surface-water runoff sufficiently to cause erosion.

This is surprising in view of the apparent dependence of people on domesticated animals and cereals, with little evidence for the use of wild animals and plants. Equipment used for tillage is likely to have been sophisticated in view of the depth of arable soils and the marks of tillage in the bedrock. Also, we are dealing with people who managed woodlands for hazel coppicing, constructed sophisticated trackways like the Sweet Track in the Somerset Levels, built grand tombs like Maes Howe in Orkney, and undertook sea voyages. Nor did their farming consist of short episodes of slash-and-burn agriculture. There was settlement and agriculture in our study area over several centuries, even if precisely the same places were not occupied continuously. The period began around 6000 years ago, and there is evidence of cereal cultivation outside the area as early as 6200 BP (Edwards and Hirons 1984); yet it is not until around 4500 BP that erosion of chalk soils began (Bell and Walker 1992).

So maybe Neolithic people were conserving soils by avoiding slopes, manuring, planting legumes among the crops or fallowing, and we know from South Street that there were at least three phases of tillage, the last two separated by grassland. They may also occasionally have moved the whole settlement and agricultural system just a few kilometres away, as suggested by the short-lived arable phase at Easton Down and the spread of bracken in some areas as indicated by pollen (Smith 1984). Alternatively it may be that at this stage in soil history, soils had a fertility and structure that could sustain the crumb structure under cultivation, especially in view of the contribution from non-calcareous drift, later eroded. Brownearth or brown calcareous soils have a crumb structure that is maintained by a balance between calcium content giving high biological activity, itself a factor in crumb formation, and mineralisation of litter which takes place more rapidly in reduced calcareous conditions but which is needed to replenish the humic components of the crumbs (Duchaufour 1982). Land may not have been a limiting factor early on in the Neolithic. Later on, the evidence for conflicts in the later stages of some causewayed camps and for developing hierarchies within Neolithic communities, perhaps related to land disputes, suggests otherwise.

There is also the possibility that earlier Neolithic soils *did* erode but no evidence remains. Thus truncation of the basal soil is widespread in valleys, with later Neolithic or later deposits abutting the subsoil (Allen 1992; Bell 1992), although this may reflect incorporation of the original soil into the later one rather than its removal. Allen and Bell suggest that water-tables were higher in the Neolithic than today and that seasonally temporary streams, called Lavants or Winterbournes and still a feature of the chalk, might have existed higher up and have been responsible for washing eroded soils right out of the system. However, there is no convincing case of earlier Neolithic sediment being preserved, even in the major river valleys. Indeed, Bell points out that alluviation in southern English river valleys is generally of medieval origin at the earliest.

Later erosion

The absence of earlier Neolithic erosion is all the more striking when we see how much erosion took place from the later Neolithic onwards (e.g. Allen 1992; Macphail 1992). By the Middle Bronze Age it was considerable, with many valleys in the chalklands containing colluvium, sometimes up to 3 m thick, from this period onwards. This may have been due to the state of the soils after the earlier Neolithic and not to differences of land-use or climate. Thus there is evidence from several sites for partial or total decalcification and loss of soil structure. The soil under the Marden henge in the Vale of Pewsey was an argillic brownearth (Wainwright 1972); the soil under the Avebury henge has a turf-line which was practically decalcified and structureless, virtually a sediment; several turf-lines within long barrow ditches show the same feature – Skendleby Giants' Hills II, Easton Down, South Street and Maiden Castle (Evans 1990). This may have been a consequence of earlier land-use and abandonment, relaxation of grazing, perhaps in a generally open land of species-poor grassland, and the prevention of scrub or woodland regeneration by a thick grass mat. A combination of low soil biological activity and a build-up of raw organic matter at the soil surface, with a loss of calcium carbonate, could have led to weakening of soil crumb structure. It only needed the vegetation to be removed for erosion to occur.

Conclusions

One of the main conclusions from the point of view of methodology to have come out of this study is that the significance of individual sequences can be greatly enhanced by reference to other sites and problems. We could hardly appreciate the significance of the Avebury sequence without considering soil profiles under earlier monuments and the whole business of soil erosion in the Neolithic and Bronze Age periods. We have fed back from the region to the site and in doing so enhanced the significance of both.

SOILS AND HUMAN LAND-USE

As we have just shown, interpretation should be done at a variety of scales and with a variety of kinds of evidence. The next three case studies (which must be considered as a group) exemplify this procedure explicitly, showing together a sequence of investigation into the relationships between soils and human use.

Terrace agriculture in semi-arid mountainous localities

First of all, at the site and local scale, there is the work of Sandor (1992) on prehistoric cultivation terraces in two semi-arid mountainous regions, New Mexico and Peru. An important methodological aspect of this work is the comparison of natural and agricultural soils. In the south-western New Mexico locality of the Sapillo Valley, areas of prehistoric cultivation terraces (AD 1000–1500) were first of all considered in their contemporary climate (tree-ring, pollen, fauna and geological data) and topographical setting. The latter was quite narrow and chosen in relation to rainfall, slope and local water supply. The terraces were placed directly downslope from unterraced soils with strong argillic horizons which favoured runoff onto the terrace fields. Next, areas of cultivated soils were compared with areas of uncultivated soils by field description, mapping and laboratory analyses of selected profiles. The cultivated soils had thicker A horizons, and were less organic, more blocky and more nearly massive in structure by comparison with soils in the uncultivated areas. Laboratory analyses suggested lower fertility (reduced N, P and organic C), weaker crumb structure and greater susceptibility to erosion. The last was also seen in the field, with gullies up to 1.2 m deep and 4 m wide in some places, cutting right down into the Bt horizons, and contrasting with the smooth surfaces of the uncultivated areas. Comparisons were also made with recent accumulations behind fallen trees on uncultivated surfaces in an attempt to identify differences between cultivated soils and soils on terraces alone. Experiments were undertaken with crop growth (using a traditional variety of corn, *Zea mays*, the main prehistoric crop) on the two types of soils, demonstrating a better yield on the uncultivated soils.

In the other study, in the Colca Valley of Peru on the semi-arid western side of the Andes, observations on modern terrace agriculture initiated and mainly used between 1300 and 1700 BP (but parts still in use today) suggested that conservation measures might maintain productivity on steep slopes and that these were used in prehistory. These traditional crop- and soil-management practices include intercropping, that is the use of two crops at once in the same field; crop rotation with nitrogen-fixing legumes and fallowing; terracing, whereby walls are set into B horizons (argillic and duripans) to minimise erosion; low-impact tillage to minimise structural damage and maintain organic matter and tilth; fertilisation with livestock manure, ashes and burnt vegetation; and irrigation by canals. Field archaeological evidence of irrigation and terracing suggested that at least some of these processes took place in the past. Soil survey over a wider area showed the distribution of fertile and poorer soils.

Soil/human relationships on a Mediterranean karst island

The two studies discussed above are about the technological relationship between agriculture and soils at a local scale. Now we go further and look at broader relations between human communities and agricultural land. At the micro-regional scale, there is the work of Gaffney and Stancic (1991) and Gaffney *et al.* (1995) on the island of Hvar off the coast of Croatia (Fig. 15.5), a karst region of Mediterranean climate. The first step was to investigate the spatial relationship between field archaeology of the Later Bronze Age/Early Iron Age and soils. This shows a positive correlation between site exploitation territories (time-distance cost surface catchments) for each of the largest defended sites (hillforts) and the best areas of agricultural land; mostly the hillforts lie at the edge of the upland areas of over 200 m. Cairns were also usually associated with fertile soils and hillfort catchments, although some concentrations of cairns and good soils were beyond the hillfort catchments (Fig. 15.5). If these cairns were the result of stone clearance for agriculture (although a funerary role as well is not excluded by this) then they suggest an arable rather than pastoral basis for farming. An initial interpretation of the major hillfort

CASE STUDIES

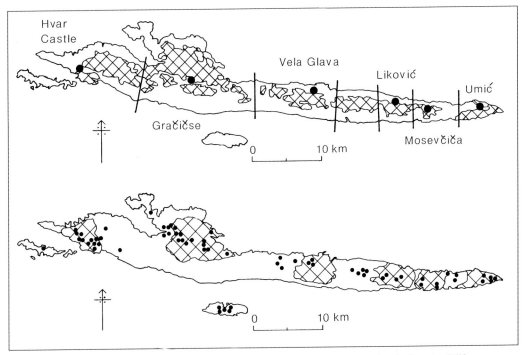

Figure 15.5 The island of Hvar, Croatia. Upper: very good and good soils cross-hatched, major hillforts as dots, and territories defined by Thiessen polygons; note that a territory based on a presumed missing hillfort has been indicated between Vela Glava and Likovic. Lower: the distribution of cairns (dots), superimposed on hillfort territories (cross-hatched) as defined by time-distance (p. 189).

distributions is that they were sited to control and possess good agricultural land. This was needed because of the threat of erosion of the limestone soils through agriculture. Ultimately a hierarchical society emerged, associated with elites whose power was based on access to resources and their control. A different approach to these distributions sees the cairns as being of primary importance in relation to access to good agricultural land. There is no clear separation between cairns associated with burial and those associated with field clearance, while the often profuse domestic rubbish within the mounds suggests purification rituals connected with the maintenance of soil fertility. So these mounds, and especially a few individual large mounds associated with the large hillforts, are seen as legitimating control and access to agricultural land by fertility rituals mediated through ancestors and the dead. The activities that took place at the hillforts, cairns and agricultural land and the relationships between them may have been a lot more complicated than their distributional relationships at first suggest.

The macro-regional view

Finally, at the macro-regional scale, is the overview of sediment and settlement sequences in Mediterranean countries by Bintliff (1992) and Wilkinson (1988). In Greece and other countries of the Mediterranean, Bintliff (1992, 126) suggested that episodes of erosion and sedimentation (largely alluviation) originated from periods of intensive human land-use. In support of this he demonstrated that there was a positive correlation of sediment sequences in different regions and that the periods of sedimentation (=erosion, =agricultural intensification) matched demographic peaks in land-use and settlement histories as identified from field survey (mainly numbers of sites as identified by pottery evidence and trace-element analysis), excavation and politico-historical sources. Wilkinson (1988) suggested a mechanism for this: the density of artefacts around settlements was partly related to a decrease in animal manure, which was being used for fuel as woodland was destroyed, and its replacement by domestic refuse containing large quantities of pottery. The evidence is circumstantial, and even if erosion is related to human settlement, there is still the matter of finer timing. Thus the actual period of erosion could be during the initial period of land clearance and agriculture, especially if the soil is in a fragile state through, say, senescence or woodland grazing; it could be during the most intensive period of land-use, due to either grazing or tillage; it could have taken place during economic downturn where conservation measures like terrace maintenance, manuring and crop rotation were neglected; or it could have been at total abandonment. Bintliff proposed that the main period of erosion was at the disjunction points between phases of unusually dense population and large-scale depopulation; Pope and van Andel (1984) suggested more or less the same, namely at periods of economic downturn; and Wilkinson's work also suggests timing late in the relevant period of land-use.

Erosion and deposition events were thus linked with expansion and contraction cycles of human population, suggesting a causal relationship. However, as Bintliff hinted (1992, 127) and later showed (1997), there are complications in the simplistic assumption that settlement density is equivalent to agricultural intensity. This is especially the case where one region is dependent on another for some of its key resources, especially those related to land-use and woodland like grain, timber and metalwork, the last needing timber for its production. The highly populated southern states of climax Classical Greece, for example, imported much of their timber and grain from the far less densely populated states of Macedonia in the north, so one might expect in this case that in both of these areas the pattern of positive correlation of periods of high population and periods of soil erosion would be inverted.

This is all very simplistic. Tectonic and sea-level changes and climate have not been mentioned, nor have short-term extreme weather events, which are increasingly seen as causal in depositional episodes, and the contingencies of history have been unaccounted for. The main aim of these examples has been to show how different kinds and scales of data can be used to suggest relationships between soil erosion and human activities.

These three examples have been placed close together, and not considered in detail, to bring out one very important point: quite different approaches can be made to address essentially the same questions, namely how soils change and how sediments form, especially in relation to erosion down slopes. Sandor looks at detailed local aspects of soil ecology but the study would have benefitted from a look at the distributional relations of soils and settlement archaeology. Gaffney and Stancic concentrate on distributional relations of soils and settlement, and develop these in the ideational sphere, but there is no consideration of the timing of erosion events in relation to settlement and sediment histories. Bintliff and other workers in Mediterranean countries have developed detailed sequences of sedimentation/erosion histories and related these to settlement and demographic change, but the concentration on correlation, the use of circumstantial evidence in inferring the origins and causes of erosion, and the polarisation of the debate between climate and anthropogenic causes are damaging (cf. Bell and Boardman 1992; Needham and Macklin 1992; Lewin *et al.* 1995). Crop growth experiments and observations on modern practices could play a part in both the Hvar and other Mediterranean studies. None of the studies takes a really close look at the processes going on as the deposits were laid down, none is explicitly ecological in its outlook, and none uses micromorphology. What is needed is the use of all these approaches, and others besides, in one region. In particular there needs to be an open-mindedness with regard to the study of soil and sediment sequences in terms of causes, the identification of what was actually happening in ecological terms as these formed, and detailed soil/archaeology distribution studies before any inferences are made with regard to overall causes.

Strong evidence of functional or causal relationships between human activities and soil distributions or change comes from circumstantial associations. Particularly persuasive are situations where changes in vegetation like woodland clearance and burning are actually represented by pollen sequences and charcoal in the deposits we are studying, like river terraces; if the chronology of terrace formation and human settlement expansions tally then this strengthens the argument even more. The studies of Tipping (1992; Mercer and Tipping 1994) in the Cheviots and of Hunt and Gilbertson (1995) in Tuscany are particularly convincing in these respects. We can make things more precise by approaching this archaeologically, as Bintliff (1992) and Pope and van Andel (1984) have done, and look for more precise evidence of what was going on in the human communities in relation to the timing of change and what might have led them to neglect soil conservation. However, the evidence is still circumstantial. People may have been present as indicated by pottery and charcoal in eroded soils, woodland clearance may be related to agriculture as indicated by pollen of cereals and weeds of cultivation, and chronologically we may be able to match the erosion episodes with contemporary settlement evidence in which, say, intensive sheep farming was taking place, or where there is archaeological evidence for population downturn and land neglect, hinting at causes. But we are still no nearer explaining under what conditions, why and how soils changed and/or eroded because there are no evidential links between the stages of the proposed sequence other than circumstantial ones. The same applies to discussions of a whole range of environmental changes such as blanket peat formation, podsolisation, fire horizons and lake-level changes. What is needed is some kind of signature of the processes we are trying to identify. Moore (1993) comes close to this in his discussion of blanket peat formation where he gets at the processes which were taking place in the soils leading up to the start of peat formation, like weakening crumb structure, pore-clogging by charcoal and dusty clay, and waterlogging. In terms of soil erosion we need to identify what changes were taking place in the soil which led to the weakening of its crumb structure and its becoming susceptible to erosion and how these were caused, whether by senescence (which is rarely acknowledged, let alone discussed in detail), natural vegetation change, climate, volcanic ash or acid fallout, or overgrazing, and how the protective vegetation cover was removed. Micromorphology is probably the best approach because the data are directly related to the soil and its history (unlike pollen, charcoal or snails), there is a wide range of signals of specific activities and processes, and they can be preserved even in eroded soils because crumbs or soil aggregates may not be completely broken down before they erode (Imeson and Jungerius 1976).

This problem of circumstantial evidence clearly applies to the interpretation of apparent clearance events around Hadrian's Wall and Neolithic monuments in southern England, too. When there are differing interpretations of the field evidence for pre-Roman settlement, we clearly need a different form of evidence by which to address the same question, with the aim of producing an unambiguous result. In the end, this is the message of this chapter, that very different lines of investigation have to be brought to bear, and their results closely integrated, if subjective and potentially misleading coincidences are to be distinguished from real cause and effect.

Sixteen

The Human Niche: A Basic Unit of Study for Archaeology

No physical entity can be said to exist entirely within its obvious physical bounds, nor within some timeless state independent of the past. Maple trees and people, for example, each have a relationship with the wider universe beyond themselves, and with the history of their particular populations and species. Similarly, a patch of soil exists in the context of the biophysical environment around it, and of the history of pedogenic factors which have impinged upon it. This is why we have in this book made rather limited use of the terms 'landscape' and 'environment'.

In the former case, 'landscape' has a vernacular use which embodies visible landforms and their current biota, whereas one aim of this book has been to emphasise the incorporation of different time-scales and processes into past and present ecosystems, giving a more dynamic and, we admit, intangible paradigm than is generally understood by the term 'landscape'. Our avoidance of the term is idiosyncratic, but purposive. In the latter case, 'environment' is subject to too much semantic juggling to be used without careful clarification. It is a common fallacy to speak of 'the environment' as if it were something separate, with its own realm of existence. People cannot exist outside of environment, nor can they cease to experience it, at least while remaining alive. Equally, 'environment' is a concept of the human mind, usually defined as the environment of someone or something. Therefore, since neither people nor environment can exist independently of each other, archaeology must necessarily study some integration of the two. The niche concept embodies that integration, incorporating the organism and its perception of, and response to, a range of variables which may include abiotic factors, such as temperature and moisture, and biotic factors, such as predation and competition. The basic unit of study in archaeology can thus be defined as the human niche, in all its splendid diversity and versatility. Niches do not exist independently of the species which occupy them: they exist because they are occupied, and the niche of a species is what that species can make of it. This, we believe, is a more useful paradigm for archaeology than to talk of 'past environments' as if these entities had some independent existence. We could quite well talk of 'the environment of Pleistocene mammoths', but that would focus the study on those factors which we believe mammoths may have perceived and responded to, and would not constitute archaeology. However, a discussion of Late Pleistocene Alaska as an environment in which people co-existed with, and occasionally predated, mammoths, would move the topic back into the realm of archaeology.

An equally important point is the spurious distinction between the natural and the human environment. Tim Ingold (1986) sees the biophysical environment as a series of 'essences' which are appropriated by people as 'affordances' and thus brought into the human niche. A hollow in a rock is an essence; if it is big enough for humans to enter it becomes a cave, as hollows in rocks which are too small for humans to enter are not regarded by humans as caves. If the cave affords protection from weather or beasties it becomes a shelter, and thus a more complex affordance expressing a response to the perceived environment. If we light a fire in the shelter and settle down for a few days, the shelter becomes a home. An external observer would see little change in the hollow in the rock, save for a whisp of smoke, and the hollow would remain much as it was before humans entered it. Similarly, hollows in a rock bluff being investigated by an archaeologist may be regarded as caves until evidence of past human settlement is encountered; then the cave is seen as a more complex affordance and becomes a rock-shelter or settlement. Whether that affordance concurs with that of the previous occupants is quite another matter. To a family of pack-rats, meanwhile, the cave is a cave, whilst adjacent hollows, far too small for humans to enter, are home.

Taking this further, whenever humans come into contact with anything through their senses, that something is incorporated into the human niche. As soon as we experience it we change it, interpret it and draw it into our lives as an affordance. Because we cannot experience environment outside of our lives there is no such thing as a 'natural' environment. Material culture subsumes attributes of our environment, from individual plants, animals and rocks, through crops and other gathered and collected materials, to blocks of land like mountains, river valleys and plains. For the human niche, there is no separation between natural and artificial, and any research which approaches environments from this view starts from a false premise.

The boundary between the human body or the human group and environment does not end with the skin and start with the air. We extend ourselves in various activities beyond our bodies, beyond immediate contact with environment or technology to higher levels of relevance which are created from within the mind. For example, when we kick a football, we do not experience the toe of our boot against the ball or see the track of the ball, even though these are the two most obvious results of our action. We see a pattern of other people responding to the ball's position, we see a future pattern developing from this, and we plan our actions accordingly. Socially, too, we relate to each other according to various properties of ourselves, e.g. class, academic status, history and gender. This is the 'human' environment in the narrow sense of the word, the environment of other human individuals. Ingold defines sociality as the ability of individuals to understand the subjectivities of others, especially when these relate, in turn, to other individuals in the first place. And those subjectivities are a parameter of the environment as much as rainfall or vegetation.

The human niche goes along as an interaction between the present, the past, and future goals or expectations, feeding back part of the past into the present to cope with the future. This is seen in music. When a famous closing aria is sung as a separate piece, as is too often the case, it may be no different in its physical quality from when it is sung at the end of the complete opera. But to the listener there is all the difference: at the end of the opera the sequential and cumulative context of the previous five hours matters. Selection of the good bits, the tunes, from a longer musical work is an affront; it pulls them out of context and un-trains the gymnastics of the brain.

We have taken an approach in this book which we might describe as 'holistic'. Environment cannot be satisfactorily disarticulated from the social realm, and so there cannot be archaeological reports with 'environmental data'. All archaeological data pertain to the environment. There is, however, a fearsome reductionism in archaeology. The tendency to regard deposition and post-deposition as separate processes, for example, leads to an attitude which rejects a bone assemblage for study if it shows substantial 'post-depositional' modification. A decision that bones tell us about past diet is over-reductionist if we therefore ascribe low value to an assemblage which has undergone numerous stages of modification since it was part of the dietary realm. The modifications themselves are pertinent archaeological data, a part of the 'life' of the bone assemblage. Indeed, if we take a more inclusive view, such assemblages may be more informative than those which have passed straight from the meal table to deposit.

Such reductionism is all a part of human tidiness. We classify things, environments and processes. Our books of flowers and animals usually show sexually mature individuals of particular species; books on biogeography show distributions and patterns of migration of species and communities; and books about soil processes are classified according to end results such as podsolisation and specific soil types. Putting an animal in a zoo, a mineral specimen in a museum case, or presenting and interpreting an archaeological site is not just removing them from their environment but focusing on one part of former life at the expense of another (Frontispiece). The zoning of pollen diagrams usually treats periods of change and transition as the boundaries between zones of relative constancy of inferred vegetation, rather than as zones in their own right. Similarly, our numerical analytical methods (chapter 13) are designed to find patterns, not mess. It is a part of human nature, it would seem, to make categories, to classify things. The challenge which we face when we seek to investigate the niche of past and present humans is that it defies simple categorisation. We have argued that the spatial boundaries of the 'environment' of a human population cannot readily be drawn (chapter 5; chapter 8), and a simple classification by period is unsatisfactory too. Archaeology seeks to understand a complex interweaving of human perception of, and response to, a range of factors and processes which operate over diverse time and spatial scales, with contingent interaction with one another. And we seek to do that by means of survey and excavation.

It is a commonplace that excavation is destructive, but this is simplistic and presumes the existence of objective data, awaiting disinterment. An excavation is what we make of it: in effect it is creative. Every soil pit or peat core is an act of creation, generating a dataset out of dormant material. The weakness with these investigations is that creation is so constrained by the reductionist paradigm of archaeology that we make only limited use of the information. Just as people-environment niches go forward as an interaction between the present, the past and the future – a creativity based on interpretation – so our work, our human-research niche, should go on in the same way. To take an analogy, when a piece of music is played it is interpreted by us just as much as its original creation was an interpretation by the composer or artist. The performer is not a go-between from composer to audience. The performer is interpreting the composer, not just transcribing a visual score to an audible experience. At the same time, the audience – each member individually and the whole as a body – are interpreting the performance, not just listening to it. So too with archaeology: interpretation is a continuum of creativity between that of the past by ancient humans, that of the present by archaeologists, and that of the individuals and groups who will make further interpretations from the publications, and are even now doing so from this text.

Interpretation and creativity are also unique to an age, building on past interpretations and changes from one generation to the next. Had Beethoven lived in the sixteenth century, technology aside, he could not have written the same music as he did in the nineteenth century, however similarly he otherwise saw the world. Likewise the interpretation of Beethoven's music varies according to the age. Similarly, archaeological interpretation is ongoing and cumulative. Just as with the way in which the styles of particular art movements are the ways in which particular painters saw the world, and the changing views through time of a single art movement, so it is with styles of archaeological interpretation.

So research must not be over-reductionist, nor must it be unaware of the superimposed sequence of creation and interpretation. Finally, research must not be unintelligently reactive. Just as humans have created themselves a distinctive niche by their socially elaborated interactions with one another and with other attributes of their ecosystems, so archaeology can find itself a niche which incorporates those things which our research environment determines we *ought* to do, and those things which we decide *a priori* need to be done. In our view, there is a clear case for investigating past human populations and their interactions with their ecosystems: indeed, we would argue that this *is* the proper study of archaeology.

REFERENCES

Addyman, P.V. 1982. 'The archaeologist's desiderata' in A.R. Hall and H.K. Kenward (eds), *Environmental archaeology in the urban context*. London, Council for British Archaeology, 1–5

Alcock, L. 1971. *Arthur's Britain*. London, Allen Lane

Allen, M.J. 1992. 'Products of erosion and the prehistoric land-use of the Wessex chalk' in M.G. Bell and J. Boardman (eds), *Past and present soil erosion: archaeological and geographical perspectives*. Oxford, Oxbow Books, 37–52

Andresen, J., Madsen, T., and Scollar, I. (eds) 1993. *Computing the past: quantitative methods in archaeology*. Aarhus, Aarhus University Press

Andrewartha, H.G. and Birch, L.C. 1954. *The distribution and abundance of animals*. Chicago, Chicago University Press

Andrews, P. 1990. *Owls, caves and fossils*. London, Natural History Museum

Ashbee, P., Bell, M.G., and Proudfoot, E. 1989. *The Wilsford shaft: excavations 1960–62*. London, English Heritage

Ashworth, A.C., Buckland, P.C., and Sadler, J.P. 1997. *An inordinate fondness for beetles. Studies in Quaternary entomology*. Chichester, John Wiley

Aston, M. 1985. *Interpreting the landscape: landscape archaeology and local history*. London, Routledge

Atkinson, T.C., Briffa, K.R., and Coope, G.R. 1987. 'Seasonal temperatures in Britain during the past 22,000 years, reconstructed using beetle remains', *Nature* 325, 587–92

Attwell, M.R. and Fletcher, M. 1987. 'An analytical technique for investigating spatial relationships', *Journal of Archaeological Science* 14, 1–11

Aufderheide, A.C. 1989. 'Chemical analysis of skeletal remains' in M.Y. Iscan and K.A.R. Kennedy (eds), *Reconstruction of life from the skeleton*. New York, Alan Liss, 237–60

Avery, B.W. 1973. 'Soil classification in the soil survey of England and Wales', *Journal of Soil Science* 24, 324–38

Avery, B.W. 1990. *Soils of the British Isles*. Wallingford, Oxon., CAB International

Bailey, G.N. and Davidson, I. 1983. 'Site exploitation territories and topography: two case studies from Palaeolithic Spain', *Journal of Archaeological Science* 10, 87–115

Baillie, M.G. 1993. 'Great oaks from little acorns . . . : precision and accuracy in Irish dendrochronology', in F.M. Chambers (ed.), *Climate change and human impact on the landscape*. London, Chapman and Hall, 33–41

Baillie, M.G. 1994. 'Dendrochronology raises questions about the nature of the AD 536 dust-veil event', *The Holocene* 4, 212–18

Baillie, M.G. 1995. *A slice through time*. London, Batsford

Bakker, J.A. and Groenman-van Waateringe, W. 1988. 'Megaliths, soils and vegetation on the Drenthe plateau' in W. Groenman-van Waateringe and M. Robinson (eds), *Man-made soils*. Oxford, British Archaeological Reports, 143–81

Ballantyne, C.K. 1991. 'Late Holocene erosion in upland Britain: climatic deterioration or human influence?', *The Holocene* 1, 81–5

Barber, K.E. 1988. 'A critical review of the role of pollen-analytical research in the environmental archaeology of central southern England', *Circaea* 5 (2), 111–14

REFERENCES

Battarbee, R.W. 1988. 'The use of diatom analysis in archaeology: a review', *Journal of Archaeological Science* 15, 621–44

Begon, M., Harper, J.L., and Townsend, C.R. 1990. *Ecology*, 2nd edn. Oxford, Blackwell

Behrensmeyer, A.K. and Boaz, D.E.D. 1980. 'The recent bones of Amboseli Park, Kenya, in relation to East African paleoecology' in A.K. Behrensmeyer and A.P. Hill (eds), *Fossils in the making*. Chicago, University of Chicago Press, 72–92

Bell, M.G. 1983. 'Valley sediments as evidence of prehistoric land-use on the South Downs', *Proceedings of the Prehistoric Society* 49, 119–50

Bell, M.G. 1990. 'Sedimentation rates in the primary fills of chalk-cut features' in D.E. Robinson (ed.), *Experimentation and reconstruction in environmental archaeology*. Oxford, Oxbow Books, 237–48

Bell, M.G. 1992. 'The prehistory of soil erosion' in M.G. Bell and J. Boardman (eds), *Past and present soil erosion: archaeological and geographical perspectives*. Oxford, Oxbow Books, 21–35

Bell, M. and Boardman, J. (eds) 1992. *Past and present soil erosion: archaeological and geographical perspectives*. Oxford, Oxbow Books

Bell, M.G. and Walker, M.J.C. 1992. *Late Quaternary environmental change*. Longman, Harlow

Bell, M., Fowler, P.J., and Hillson, S. (eds) 1996. *The experimental earthwork project 1960–1992*. York, Council for British Archaeology

Bellomo, R.V. 1993. 'A methodological approach for identifying archaeological evidence of fire resulting from human activities', *Journal of Archaeological Science* 20, 525–53

Benecke, N. 1987. 'Studies on early dog remains from northern Europe', *Journal of Archaeological Science* 14, 31–49

Benecke, N. 1988. *Archaeozoologische Untersuchungen an Tierknochen aus der frühmittelalterlichen Siedlung von Menzlin*. Schwerin, Museum für Ur- und Frühgeschichte

Benson, D. and Miles, D. 1974. *The upper Thames valleys: an archaeological survey of the river gravels*. Oxford, Oxford Archaeological Unit

Benson, D.G., Evans, J.G., and Williams, G. 1991. 'Excavations at Stackpole Warren, Dyfed', *Proceedings of the Prehistoric Society* 56, 179–245

Berglund, B.E. (ed.) 1986. *Handbook of Holocene palaeoecology and palaeohydrology*. Chichester, John Wiley

Bescoby, D.J. 1997. *Landscape modelling using GIS: a study of late prehistoric land-use in the Malham area*. Unpublished B.Sc. dissertation, University of Bradford

Bettis III, E.A. 1992. 'Soil morphologic properties and weathering zone characteristics as age indicators in Holocene alluvium in the Upper Midwest' in V.T. Holliday (ed.), *Soils in archaeology: landscape evolution and human occupation*. Washington, Smithsonian Institution Press, 119–44

Binford, L.R. 1978. *Nunamiut ethnoarchaeology*. New York, Academic Press

Bintliff, J. 1981. 'Archaeology and the Holocene evolution of coastal plains in the Aegean and circum-Mediterranean' in D. Brothwell and G. Dimbleby (eds), *Environmental aspects of coasts and islands*. Oxford, British Archaeological Reports, 11–31

Bintliff, J. 1985. 'The Boeotia survey' in S. Macready and F.H. Thompson (eds), *Archaeological field survey in Britain and abroad*. London, Society of Antiquaries, 196–216

Bintliff, J. 1992. 'Erosion in the Mediterranean lands: a reconsideration of pattern, process and methodology' in M.G. Bell and J. Boardman (eds), *Past and present soil erosion: archaeological and geographical perspectives*. Oxford, Oxbow Books, 125–31

Bintliff, J. 1997. 'Regional survey, demography and the rise of complex societies in the ancient Aegean: core-periphery, neo-Malthusian and other interpretive models', *Journal of Field Archaeology* 24 (1), 1–38

Bintliff, J.L., Davidson, D.A., and Grant, E.G. (eds) 1988. *Conceptual issues in environmental archaeology*. Edinburgh, Edinburgh University Press

Bird-David, N. 1992. 'Beyond "the hunting and gathering mode of subsistence": culture-sensitive observations on the Nayaka and other modern hunter-gatherers', *Man* 27, 19–44

Bishop, M.J. 1981. 'Quantitative studies on some living British wetland mollusc faunas', *Biological Journal of the Linnaean Society* 15, 299–326

Blumenschine, R.J. 1988. 'An experimental model of the timing of hominid and carnivore influence on archaeological bone assemblages', *Journal of Archaeological Science* 15, 483–502

REFERENCES

Bond, J.M. and O'Connor, T.P. 1998. *Bones from medieval deposits at 16–22 Coppergate and other sites in York*. York, Council for British Archaeology

Bordes, F. 1972. *A tale of two caves*. London, Harper Row

Bottema, S. and Ottaway, B.S. 1982. 'Botanical, malacological and archaeological zonation of settlement deposits at Gomolava', *Journal of Archaeological Science* 9, 221–46

Boule, M. 1923. *Fossil Men: elements of human palaeontology*. Edinburgh, Oliver & Boyd

Bowen, D.Q. 1978. *Quaternary geology: a stratigraphic framework for multidisciplinary work*. Oxford, Pergamon Press

Boycott, A.E. 1934. 'The habitats of land mollusca in Britain', *Journal of Ecology* 22, 1–38

Boycott, A.E. 1936. 'The habitats of fresh-water mollusca in Britain', *Journal of Animal Ecology* 5, 116–86

Boyd, W. 1990. 'Towards a conceptual framework for environmental archaeology: environmental archaeology as a key to past geographies', *Circaea* 7(2), 63–6

Bradshaw, R.H.W. 1991. 'Spatial scale in the pollen record' in D.R. Harris and K.D. Thomas (eds), *Modelling ecological change*. London, University College, 41–52

Braidwood, R.J. 1960. 'The agricultural revolution', *Scientific American* 203, 130–48

Brain, C.K. 1981. *The hunters or the hunted? An introduction to African cave taphonomy*. Chicago, Chicago University Press

Brain, C.K. and Brain, V. 1977. 'Microfaunal remains from Mirabib: some evidence of palaeo-ecological changes in the Namib', *Madoqua* 10(4), 285–93

Briggs, D.J., Gilbertson, D.D., and Harris, A.L. 1985. 'Molluscan taphonomy: the development of a predictive model and its application to the Summerton-Radley terrace of the upper Thames basin' in N.R.J. Fieller, D.D. Gilbertson and N.G. Ralph (eds), *Palaeoenvironmental investigations: research design, methods and data analysis*. Oxford, British Archaeological Reports, 67–91

Briggs, D.J., Gilbertson, D.D., and Harris, A.L. 1990. 'Molluscan taphonomy in a braided river environment and its implications for studies of Quaternary cold-stage river deposits', *Journal of Biogeography* 17, 623–37

Brothwell, D.R. and Dimbleby, G.W. (eds) 1981. *Environmental aspects of coasts and islands*. Oxford, British Archaeological Reports

Brown, A. 1987. *Fieldwork for archaeologists and local historians*. London, Batsford

Brown, A.G. 1992. 'Slope erosion and colluviation at the floodplain edge' in M.G. Bell and J. Boardman (eds), *Past and present soil erosion: archaeological and geographical perspectives*. Oxford, Oxbow Books, 77–87

Brown, A.G. 1997. *Alluvial geoarchaeology*. Cambridge, Cambridge University Press

Brown, A.G. and Keough, M.K. 1992. 'Palaeochannels and palaeolandsurfaces: the geoarchaeological potential of some Midland floodplains' in S. Needham and M.G. Macklin (eds), *Alluvial archaeology in Britain*. Oxford, Oxbow Books, 185–96

Brown, D.A. 1984. 'Prospects and limits of a phytolith key for grasses in the central United States', *Journal of Archaeological Science* 11, 345–68

Bryce, D. 1962. 'Chiroptera (Diptera) from freshwater sediments, with special reference to Malham Tarn (Yorks.)', *Transactions of the Society of British Entomologists* 15, 41–54

Buck, C.E., Cavanagh, W.G., and Litton, C.D. 1996. *Bayesian approach to interpreting archaeological data*. Chichester, John Wiley

Buckland, P.C. 1976. *The environmental evidence from the Church Street Roman sewer system*. London, Council for British Archaeology

Buckland, P.C., Sadler, J.P., and Sveinbjarnarsdottir, G. 1992. 'Palaeoecological investigations at Reykholt, western Iceland' in C.D. Morris and D.J. Rackham (eds.), *Norse and later settlement and subsistence in the North Atlantic*. Glasgow, University of Glasgow Press, 149–68

Budyko, M.I. 1982. *The earth's climate: past and future*. New York, Academic Press

Burnham, C.P. 1980. 'The soils of England and Wales', *Field Studies* 5, 349–63

Burrin, P.J. and Scaife, R.G. 1984. 'Aspects of Holocene valley sedimentation and floodplain development in southern England', *Proceedings of the Geologists' Association* 95, 81–96

REFERENCES

Bush, M.B. 1988a. 'The use of multivariate analysis and modern analogue sites as an aid to the interpretation of data from fossil mollusc assemblages', *Journal of Biogeography* 15, 849–61

Bush, M.B. 1988b. 'Early Mesolithic disturbance: a force on the landscape', *Journal of Archaeological Science* 15, 453–62

Butterworth, C.A. and Lobb, S.J. 1992. *Excavations in the Burghfield area, Berkshire*. Salisbury, Wessex Archaeology

Butzer, K.W. 1971. *Environment and archaeology: an ecological approach to prehistory*. Chicago, Aldine

Butzer, K.W. 1976. *Early hydraulic civilisation in Egypt*. Chicago, Chicago University Press

Butzer, K.W. 1982. *Archaeology as human ecology. Method and theory for a contextual approach*. Cambridge, Cambridge University Press

Byers, D.S. 1967. *The prehistory of the Tehuacán Valley. Volume 1 Environment and subsistence*. Austin, University of Texas Press

Cain, A.J. 1977. 'Variation in the spire index of some coiled gastropod shells, and its evolutionary significance', *Philosophical Transactions of the Royal Society* B 277, 377–428

Cain, A.J. 1978. 'Variation of terrestrial gastropods in the Philippines in relation to shell shape and size', *Journal of Conchology* 29, 239–45

Calow, P. 1981. 'Adaptational aspects of growth and reproduction in *Lymnaea peregra* (Gastropoda: Pulmonata) from exposed and sheltered aquatic habitats', *Malacologia* 21, 5–13

Cameron, C.M. and Tomka, S.A. (eds) 1993. *Abandonment of settlements and regions: ethnoarchaeological and archaeological approaches*. Cambridge: Cambridge University Press

Cameron, R.A.D. and Morgan-Huws, D.I. 1975. 'Snail faunas in the early stages of a chalk grassland succession', *Biological Journal of the Linnean Society* 7, 215–29

Campbell, E. and Lane, A. 1989. 'Llangorse: a 10th-century royal crannog in Wales', *Antiquity* 63, 675–81

Carss, D.N. and Parkinson, S.G. 1996. 'Errors associated with otter *Lutra lutra* faecal analysis. I. Assessing general diet from spraints', *Journal of Zoology* 238(2), 301–18

Carter, S. 1990a. 'The distribution of the land snail *Vitrea contracta* (Westerlund) in a calcareous soil on Martin Down, Hampshire, England', *Circaea* 7, 91–3

Carter, S. 1990b. 'The stratification and taphonomy of shells in calcareous soils: implications for land snail analysis in archaeology', *Journal of Archaeological Science* 17, 495–507

Caseldine, A. 1990. *Environmental archaeology in Wales*. Lampeter, St Davids University College

Chambers, F.M. 1978. 'A radiocarbon-dated pollen diagram from Valley Bog, on the Moor House National Nature Reserve', *New Phytologist* 80, 273–80

Chambers, F.M. 1993a. 'Late-Quaternary change and human impact: commentary and conclusions' in F.M. Chambers (ed.), *Climate change and human impact on the landscape*. London, Chapman and Hall, 247–59

Chambers, F.M. (ed.) 1993b. *Climate change and human impact on the landscape*. London, Chapman and Hall

Chambers, F.M. and Price, S.-M. 1988. 'The environmental setting of Erw-wen and Moel-y-Gerddi: prehistoric enclosures in upland Ardudwy, North Wales', *Proceedings of the Prehistoric Society* 54, 93–100

Cherry, J.F., Gamble, C., and Shennan, S. (eds) 1978. *Sampling in contemporary British archaeology*. Oxford, British Archaeological Reports

Child, A.M. 1995. 'Microbial taphonomy of archaeological bone', *Studies in Conservation* 40, 19–30

Childe, V.G. 1936. *Man makes himself*. London, Watts & Co.

Clark, A. 1990. *Seeing beneath the soil*. London, Batsford

Clark, J.G.D. 1952. *Prehistoric Europe: the economic basis*. London, Methuen

Clark, J.G.D. 1989. *Economic archaeology*. Cambridge, Cambridge University Press

Clason, A.T. and Prummel, W. 1977. 'Collecting, sieving and archaeozoological research', *Journal of Archaeological Science* 4, 171–5

Coles, G.M. 1990. 'A note on the systematic recording of organic-walled microfossils (other than pollen and spores) found in archaeological and Quaternary palynological preparations', *Circaea* 7, 103–11

Coles, J. 1973. *Archaeology by experiment*. London, Hutchinson

Colinvaux, P. 1980. *Why big fierce animals are rare*. London, Penguin Books
Colinvaux, P. 1982. 'Towards a theory of history: fitness, niche and clutch size of *Homo sapiens*', *Journal of Ecology* 70, 393–412
Colinvaux, P. 1993. *Ecology 2*. New York, John Wiley
Cong, S. and Ashworth, A.C. 1997. 'The use of Correspondence Analysis in the analysis of fossil beetle assemblages' in A.C. Ashworth, P.C. Buckland and J.P. Sadler (eds), *An inordinate fondness for beetles. Studies in Quaternary entomology*. Chichester, John Wiley
Conway, J.S. 1983. 'An investigation of soil phosphorus distribution within occupation deposits from a Romano-British hut group', *Journal of Archaeological Science* 10, 117–28
Coope, G.R. 1977. 'Fossil coleopteran assemblages as sensitive indicators of climatic changes during the Devensian (last) cold stage', *Philosophical Transactions of the Royal Society* B 280, 313–48
Coope, G.R. and Brophy, J.A. 1972. 'Late Glacial environmental changes indicated by a coleopteran succession from North Wales', *Boreas* 1, 97–142
Corbet, G.B. and Southern, H.N. 1977. *The handbook of British mammals*, 2nd edn. Oxford, Blackwell
Cornwall, I.W. 1958. *Soils for the archaeologist*. London, Phoenix House
Cornwall, I.W. 1964. *Bones for the archaeologist*. London, Phoenix House
Courty, M.-A. 1992. 'Soil micromorphology in archaeology', *Proceedings of the British Academy* 77, 39–59
Courty, M.-A., Goldberg, P., and Macphail, R. 1989. *Soils and micromorphology in archaeology*. Cambridge, Cambridge University Press
Craddock, P.T., Gurney, D., Pryor, F., and Hughes, M.J. 1985. 'The application of phosphate analysis to the location and interpretation of archaeological sites', *Archaeological Journal* 142, 361–76
Cronberg, G. 1986. 'Blue-green algae, green algae and Chrysophyceae in sediments' in B.E. Berglund (ed.), *Handbook of Holocene palaeoecology and palaeohydrology*. Chichester, John Wiley, 507–26
Cunliffe, B.W. 1983. *Danebury: anatomy of an Iron Age hillfort*. London, Batsford
Davidson, D.A. 1976. 'Processes of tell formation and erosion' in D.A. Davidson and M.L. Shackley (eds), *Geoarchaeology: earth science and the past*. London, Duckworth, 255–66
Davidson, D.A. 1980. *Soils and land-use planning*. New York, Longman
Davidson, D.A. and Simpson, I.A. 1984. 'The formation of deep topsoils in Orkney', *Earth Surface Processes and Landforms* 9, 75–81
Davies, P. 1998. 'Numerical analysis of subfossil wet ground molluscan taxocenes from overbank alluvium at Kingsmead Bridge, Wiltshire', *Journal of Archaeological Science* 25, 39–52
Davies, W. 1988. *Small worlds: the village community in early Medieval Brittany*. London, Duckworth
Davies, W. and Astill, G. 1994. *The east Brittany survey*. Aldershot, Scolar Press
Davis, S.J.M. 1987. *The archaeology of animals*. London, Batsford
de Geer, G. 1910. *A geochronology of the last 12000 years*. Stockholm, International Geological Congress
de Lumley, H. 1972. *La Grotte de l'Hortus*. Marseilles, Centre de la Recherche Scientifique
Deevey, E.S. 1949. 'Biogeography of the Pleistocene', *Bulletin of the Geological Society of America* 60 (9), 1315–16
Deith, M.R. 1983. 'Molluscan calendars: the use of growth-line analysis to establish seasonality of shellfish collection at the Mesolithic site of Morton, Fife', *Journal of Archaeological Science* 10, 423–40
Deith, M.R. 1985. 'Seasonality from shells: an evaluation of two techniques for seasonal dating of marine molluscs' in N.R.J. Fieller, D.D. Gilbertson and N.G.A. Ralph (eds), *Palaeobiological investigations: research design, methods and data analysis*. Oxford, British Archaeological Reports, 119–30
Deith, M.R. 1990. 'Clams and salmonberries: interpreting seasonality data from shells' in C. Bonsall (ed.), *The Mesolithic in Europe*. Edinburgh, Edinburgh University Press, 73–9
Delcourt, P.A. and Delcourt, H.R. 1980. 'Pollen preservation and Quaternary environmental history in the south-eastern United States', *Palynology* 4, 215–31
Dennell, R. 1997. 'The worlds oldest spears', *Nature* 385, 767–8
Devoy, R.J.N. 1979. 'Flandrian sea-level changes and vegetational history of the lower Thames estuary', *Philosophical Transactions of the Royal Society* B 285, 355–410
Diamond, J. 1973. 'Distributional ecology of New Guinea birds', *Science* 179, 759–69

REFERENCES

Dickson, C. 1987. 'The identification of cereals from ancient bran fragments', *Circaea* 7(2), 103–11

Dimbleby, G.W. 1962. *The development of British heathlands and their soils*. Oxford, Clarendon Press

Dimbleby, G.W. 1976. 'Climate, soil and man', *Philosophical Transactions of the Royal Society of London* B 275, 197–208

Dimbleby, G.W. 1978. *Plants and archaeology*. London, John Baker

Dimbleby, G.W. 1985. *The palynology of archaeological sites*. London, Academic Press

Dockrill, S.J., Bond, J.M., and O'Connor, T.P. 1998. 'Beyond the burnt mound: the South Nesting palaeolandscape project' in V. Turner (ed.), *The shaping of Shetland*. Lerwick, The Shetland Times, 61–82

Driver, J.C. 1992. 'Identification, classification and zooarchaeology', *Circaea* 9, 35–47

Duchaufour, P. 1982. *Pedology: pedogenesis and classification*. London, George Allen & Unwin

Dumayne, L. and Barber, K.E. 1994. 'The impact of the Romans on the environment of northern England: pollen data from three sites close to Hadrian's Wall', *The Holocene* 4, 165–73

Dumayne-Peaty, L. and Barber, K.E. 1997. 'Archaeological and environmental evidence for Roman impact on vegetation near Carlisle, Cumbria: a comment on McCarthy', *The Holocene* 7, 243–46

Duplessy, J.-C. and Overpeck, J. (eds) 1996. *The PAGES/CLIVAR intersection*. Venice, IGBP-WCRP

Edwards, K.J. and Hirons, K.R. 1984. 'Cereal pollen grains in pre-elm decline deposits: implications for the earliest agriculture in Britain and Ireland', *Journal of Archaeological Science* 11, 71–80

Efremov, J.A. 1940. 'Taphonomy: new branch of palaeontology', *Pan-American Geologist* 74, 81–93

Ellenberg, H. 1988. *Vegetation ecology of Central Europe*. Cambridge, Cambridge University Press

Ellison, A. 1980. 'Towards a socio-economic model for the Middle Bronze Age in southern England' in I. Hodder, G. Isaac and N. Hammond (eds), *Pattern of the past. Studies in honour of David Clarke*. Cambridge, Cambridge University Press, 413–38

Ellison, A. and Harriss, J. 1972. 'Settlement and land-use in the prehistory and early history of southern England: a study based on locational methods' in D.L. Clarke (ed.), *Models in archaeology*. London, Methuen, 911–62

Elton, C.S. 1927. *Animal ecology*. New York, Macmillan

Enckell, P.H. and Rundgren, S. 1988. 'Anthropochorous earthworms (Lumbricide) as indicators of abandoned settlements in the Faroe Islands', *Journal of Archaeological Science* 15, 439–52

Evans, D.H. and Tomlinson, D.G. 1992. *Excavations at 33–35 Eastgate, Beverley 1983–86*. Sheffield, Sheffield Excavation Reports

Evans, J.G. 1972. *Land snails in archaeology*. London, Academic Press

Evans, J.G. 1990. 'Notes on some Late Neolithic and Bronze Age events in long barrow ditches in southern and eastern England', *Proceedings of the Prehistoric Society* 56, 111–16

Evans, J.G. 1991. 'An approach to the interpretation of dry-ground and wet-ground molluscan taxocenes from central-southern England' in D.R. Harris and K.D. Thomas (eds), *Modelling ecological change*. London, University College, 80–95

Evans, J.G. 1993. 'The influence of human communities on the English chalklands from the mesolithic to the Iron Age: the molluscan evidence' in F.M. Chambers (ed.), *Climate change and human impact on the landscape*. London, Chapman and Hall, 147–56

Evans, J.G. and Jones, H. 1973. 'Subfossil and modern land-snail faunas from rock-rubble habitats', *Journal of Conchology* 27, 103–29

Evans, J.G. and Rouse, A.J. 1990. 'Small-vertebrate and molluscan analysis from the same site', *Circaea* 8, 75–84

Evans, J.G. and Simpson, D.D.A. 1991. 'Giants' Hills 2 long barrow, Skendleby, Lincolnshire', *Archaeologia* 109, 1–45

Evans, J.G. and Smith, I.F. 1983. 'Excavations at Cherhill, north Wiltshire, 1967', *Proceedings of the Prehistoric Society* 49, 43–117

Evans, J.G. and Williams, D. 1991. 'Land Mollusca from the M3 archaeological sites: a review' in P.J. Fasham and R.J.B. Whinney (eds), *Archaeology and the M3*. Winchester, Hampshire Field Club, 113–42

Evans, J.G., Limbrey, S., Máté, I., and Mount, R. 1993. 'An environmental history of the Upper Kennet valley, Wiltshire, for the last 10,000 years', *Proceedings of the Prehistoric Society* 59, 139–95

Evans, J.G., Pitts, M.W., and Williams, D. 1985. 'An excavation at Avebury, Wiltshire, 1982', *Proceedings of the Prehistoric Society* 51, 305–10

Everitt, B. 1980. *Cluster analysis*, 2nd edn. London, Heinemann Educational

Faegri, K. and Iversen, J. 1989. *Textbook of pollen analysis*, 4th edn. Copenhagen, Munksgaard

Fagan, B.M. 1991. *Ancient North America*. London, Thames and Hudson

FAO 1974. *FAO-UNESCO soil map of the world*. Paris, UNESCO

Ferring, C.R. 1992. 'Alluvial pedology and geoarchaeological research' in V.T. Holliday (ed.), *Soils in archaeology: landscape evolution and human occupation*. Washington, Smithsonian Institution Press, 145–67

Fisher, P.F. 1982. 'A review of lessivage and Neolithic cultivation', *Journal of Archaeological Science* 9, 299–304

Fisher, R.A., Corbet, A.S., and Williams, C.B. 1943. 'The relationship between the number of species and the number of individuals in a random sample of an animal population', *Journal of Animal Ecology* 12, 42–58

Fishpool, M. 1992. *Investigation of the responses of land snails and carabid beetles to vegetation boundaries, with reference to the interpretation of subfossil assemblages*. Unpublished Ph.D. thesis, University of Cardiff

Fleming, A. and Ralph, N. 1982. 'Medieval settlement and land use on Holne Moor, Dartmoor: the landscape evidence', *Medieval Archaeology* 26, 101–37

Foster, R.J. 1985. *Geology*, 5th edn. London, Charles Merrill

French, D. 1973. 'Asvan 1968–1972: an interim report', *Anatolian Studies* 23, 71–307

Gaffney, C.F. and Gaffney, V.L. 1988. 'Some quantitative approaches to site territory and land use from the surface record' in J.L. Bintliff, D.A. Davidson and E.G. Grant (eds), *Conceptual issues in environmental archaeology*. Edinburgh, University Press, 82–90

Gaffney, V. and Stancic, Z. 1991. *GIS approaches to regional analysis: a case study of the island of Hvar*. Ljubljana, Znanstveni institut Filozofske fakultete

Gaffney, V. and Tingle, M. 1985. 'The Maddle Farm (Berks.) project and micro-regional analysis' in S. Macready and F.H. Thompson (eds), *Archaeological field survey in Britain and abroad*. London, Society of Antiquaries of London, 67–73

Gaffney, V., Stancic, Z., and Watson, H. 1995. 'Moving from catchments to cognition: tentative steps towards a larger archaeological context for GIS', *Scottish Archaeological Review* 9 and 10, 41–64

Gamble, C.S. 1979. 'Sampling and archaeozoological studies' in M. Kubasiewicz (ed.), *Archaeozoology*. Szczecin, Agricultural Academy, 120–8

Gamble, C. 1986. *The Palaeolithic settlement of Europe*. Cambridge, Cambridge University Press

Gardiner, J.P. 1980. 'Land and social status – a case study from eastern England' in J. Barrett and R. Bradley (eds), *Settlement and society in the later British Bronze Age, pt. i*. Oxford, British Archaeological Reports, 101–14

Gardiner, N.P. 1991. *Small-scale distributions of two modern land-snail faunas: islands and boundaries of relevance to the interpretation of subfossil assemblages*. Unpublished Ph.D. thesis, University of Cardiff

Gee, J.H.R. and Giller, P.S. 1991. 'Contemporary community ecology and environmental archaeology' in D.R. Harris and K.D. Thomas (eds), *Modelling ecological change*. London, Institute of Archaeology, 1–12

Gelling, M. 1978a. *Signposts to the past*. London, J.M. Dent & Sons

Gelling, M. 1978b. 'The effect of man on the landscape: the place-name evidence in Berkshire' in S. Limbrey and J.G. Evans (eds), *The effect of man on the landscape: the Lowland Zone*. London, Council for British Archaeology, 123–5

Gelling, M. 1984. *Place-names in the landscape*. London, J.M. Dent & Sons

Gennard, D.E. 1985. 'Observations on the evidence for flax growth in Ireland provided by pollen analysis', *Circaea* 3, 159–62

Gifford, D.P. 1980. 'Ethnoarchaeological contributions to the taphonomy of human sites' in A.K. Behrensmeyer and A.P. Hill (eds), *Fossils in the making*. Chicago, University of Chicago Press, 93–106

Gilbertson, D.D. 1995. 'Studies of lithostratigraphy and lithofacies: a selective review of research

REFERENCES

developments in the last decade and their applications to geoarchaeology' in A.J. Barham and R.I. Macphail (eds), *Archaeological sediments and soils: analysis, interpretation and management*. London, University College London, 99–144

Gilbertson, D., Kent, M., and Grattan, J. (eds) 1996. *The Outer Hebrides: the last 14,000 years*. Sheffield, Sheffield Academic Press

Godwin, H. 1956. *The history of the British flora, a factual basis for phytogeography*. Cambridge, Cambridge University Press

Godwin, H. 1975. *History of the British flora*, 2nd edn. Cambridge, Cambridge University Press

Goldberg, P. 1992. 'Micromorphology, soils and archaeological sites' in V.T. Holliday (ed.), *Soils in archaeology: landscape evolution and human occupation*. Washington, Smithsonian Institution Press, 1–39

Gould, P.R. 1963. 'Man against his environment: a game-theoretic framework', *Annals of the Association of American Geographers* 53, 290–7

Gould, R.A. 1980. *Living archaeology*. Cambridge, Cambridge University Press

Gower, J.C. 1975. 'Goodness-of-fit criteria for classification and other patterned structures', *Proceedings of the 8th Conference on Numerical Taxonomy*, 38–62

Graham, M. 1993. 'Settlement organization and residential variability among the Raramuri' in C.M. Cameron and S.A. Tomka (eds), *Abandonment of settlements and regions: ethnoarchaeological and archaeological approaches*. Cambridge, Cambridge University Press, 25–42

Graham, R.W. 1985. 'Diversity and community structure of the Late Pleistocene mammal fauna of North America', *Acta Zoologica Fennica* 170, 181–92

Gray, J.S. 1981. *The ecology of marine sediments: an introduction to the structure and function of benthic communities*. Cambridge, Cambridge University Press

Greene, F.J. and Lockyear, K. 1994. 'Seeds, sherds and samples: site formation processes at the Waitrose site, Romsey' in R. Luff and P. Rowley-Conwy (eds), *Whither environmental archaeology?* Oxford, Oxbow Books, 91–104

Greig, J. 1988. 'Some evidence of the development of grassland plant communities' in M.K. Jones (ed.), *Archaeology and the flora of the British Isles*. Oxford, Oxford University Committee for Archaeology, 39–54

Griffiths, H.I., Ringwood, V., and Evans, J.G. 1994. 'Weichselian Late-Glacial and early Holocene molluscan and ostracod sequences from lake sediments at Stellmoor, north Germany', *Archiv für Hydrobiologie, Supplement* 99(3), 357–80

Griffiths, H.I., Rouse, A., and Evans, J.G. 1993. 'Processing freshwater ostracods from archaeological deposits with a key to the valves of the major British genera', *Circaea* 10(2), 53–62

Grinnel, J. 1904. 'The origin and distribution of the chestnut-backed chicadee', *Auk* 21, 364–82

Groenman-van Waateringe, W. and Robinson, M. (eds) 1988. *Man-made soils*. Oxford, British Archaeological Reports

Gudmundsson, G. 1991. *The ethnohistory and archaeology of Lyme grass* (Elymus arenarius) *in Iceland*. Unpublished B.Sc. thesis, University College London

Hagelberg, E. 1996. 'Mitochondrial DNA in ancient and modern humans' in A.J. Boyce and C.G.N. Mascie-Taylor (eds), *Molecular biology and human diversity*. Cambridge, Cambridge University Press, 1–11

Hagelberg, E., Thomas, M.G., Cook, C.E., Sher, A.V., Baryshnikov, G.F., and Lister, A.M. 1994. 'DNA from ancient mammoth bones', *Nature* 370, 333–4

Hall, A.R. and Kenward, H.K. 1990. *Environmental evidence from the colonia*. London, Council for British Archaeology

Hall, D. 1982. *Medieval fields*. Princes Risborough, Shire Publications Ltd

Hall, V. 1989. 'A comparison of grass foliage, moss polsters and soil surfaces as pollen traps in modern pollen studies', *Circaea* 6 (1), 63–9

Halstead, P. and O'Shea, J. (eds) 1989. *Bad year economics: cultural response to risk and uncertainty*. Cambridge, Cambridge University Press

Hamblin, W.K. 1992. *Earth's dynamic systems*, 6th edn. New York, Macmillan

Hanson, C.B. 1980. 'Fluvial processes: models and experiments' in A.K. Behrensmeyer and A.P. Hill (eds), *Fossils in the making: vertebrate taphonomy and paleoecology*. Chicago, University of Chicago Press, 156–81

Harris, D.R. 1989. 'An evolutionary continuum of people–plant interaction' in D.R. Harris and G.C. Hillman (eds), *Foraging and farming*. London, Unwin Hyman, 11–26

Hastorf, C.A. and Popper, V.S. 1988. *Current palaeoethnobotany*. Chicago, Chicago University Press

Hay, R. 1976. *Geology of Olduvai Gorge: a study of sedimentation in a semi-arid basin*. Berkeley, University of California Press

Haynes, G. 1988. 'Longitudinal studies of African elephant death and bone deposits', *Journal of Archaeological Science* 15, 131–57

Hedges, R.E.M. and van Klinken, G.-J. (eds) 1995. *Special issue of Journal of Archaeological Science on bone diagenesis* 22, 145–340

Hemphill-Haley, E. 1996. 'Diatoms as an aid in identifying late-Holocene tsunami deposits', *The Holocene* 6, 439–48

Hesse, B. and Wapnish, P. 1985. *Animal bone archaeology from objectives to analysis*. Washington, Taraxacum Press

Hicks, S., Latalowa, M., Ammann, B., Pardoe, H., and Tinsley, H. 1996. *European pollen monitoring programme*. Oulu, Inqua

Higgs, E.S. (ed.) 1972. *Papers in economic prehistory*. Cambridge, Cambridge University Press

Hill, J.D. 1995. *Ritual and rubbish in the Iron Age of Wessex*. Oxford, British Archaeological Reports

Hillman, G. 1981. 'Reconstructing crop husbandry practices from charred remains of crops' in R. Mercer (ed.), *Farming practice in British prehistory*. Edinburgh, Edinburgh University Press, 123–62

Hillman, G. 1991. 'Phytosociology and ancient weed floras: taking account of taphonomy and changes in cultivation methods' in D.R. Harris and K.D. Thomas (eds), *Modelling ecological change*. London, Institute of Archaeology, 27–40

Hillman, G.C., Mason, S., de Moulins, D., and Nesbitt, M. 1996. 'Identification of archaeological remains of wheat: the 1992 London workshop', *Circaea* 12, 195–209

Hingley, R. 1996. 'Ancestors and identity in the later prehistory of Atlantic Scotland: the reuse and reinvention of Neolithic monuments and material culture', *World Archaeology* 28, 231–43

Hodder, I. 1982a. *The present past: an introduction to anthropology for archaeologists*. London, Batsford

Hodder, I. 1982b. *Symbols in action*. Cambridge, Cambridge University Press

Hodder, I. and Orton, C. 1979. *Spatial analysis in archaeology*. Cambridge, Cambridge University Press

Hodges, R. 1991. *Wall-to-wall history: the story of Roystone Grange*. London, Duckworth

Hoffman, W. 1986. 'Chironomid analysis' in B.E. Berglund (ed.), *Handbook of Holocene palaeoecology and palaeohydrology*. Chichester, John Wiley, 715–27

Hoganson, J.W. and Ashworth, A.C. 1992. 'Fossil beetle evidence for climatic change 18,000–10,000 years BP in South-Central Chile', *Quaternary Research* 37, 101–16

Holden, T. 1990. 'Transverse cell patterns of wheat and rye bran and their variation over the surface of a single grain', *Circaea* 6, 97–104

Holliday, V.T. 1992. 'Soil formation, time and archaeology' in V.T. Holliday (ed.), *Soils and archaeology*. Washington, Smithsonian Institution, 101–17

Hunt, C.O. and Gilbertson, D.D. 1995. 'Human activity, landscape change and valley alluviation in the Feccia valley, Tuscany, Italy' in J. Lewin, M.G. Macklin and J.C. Woodward (eds), *Mediterranean Quaternary river environments*. Rotterdam, A.A. Balkema, 167–76

Huntley, J.P. and Stallibrass, S. 1995. *Plant and vertebrate remains from archaeological sites in northern England: data reviews and other directions*. Unpublished report, University of Durham

Hutchinson, G.E. 1978. *An introduction to popular ecology*. New Haven, Yale University Press

Imeson, A.C. and Jungerius, P.D. 1976. 'Aggregate stability and colluviation in the Luxembourg Ardennes: an experimental and micromorphological study', *Earth Surface Processes* 1, 259–71

Ingold, T. 1986. *The appropriation of nature*. Manchester, Manchester University Press

Iversen, J. 1941. 'Landnam i Danmarks Stenalder. En pollenanalytisk Undersøgelse over det forste Landbrugs Indviskning paa Vegetationsudvikilngen', *Danmarks Geologiske Undersøgelse* RII, no. 67, 7–68

REFERENCES

Jacobsen, G.L. and Bradshaw, R.H.L. 1981. 'The selection of sites for palaeo-vegetational studies', *Quaternary Research* 16, 80–96

Jacobsen, L. and Hansen, H.-M. 1996. 'Analysis of otter (*Lutra lutra*) spraints. Part 1: comparison of methods to estimate prey proportions. Part 2: estimation of the size of prey fish', *Journal of Zoology* 238 (1), 167–80

Janssen, C.R. 1966. 'Recent pollen spectra from the deciduous and coniferous-deciduous forests of northeastern Minnesota: a study in pollen dispersal', *Ecology* 47, 804–25

Johnson, N. 1985. 'Archaeological field survey: a Cornish perspective' in S. Macready and F.H. Thompson (eds), *Archaeological field survey in Britain and abroad*. London, Society of Antiquaries of London, 51–66

Jones, A.K.G. 1985. 'Trichurid ova in archaeological deposits: their value as indicators of ancient faeces' in N.R.J. Fieller, D.D. Gilbertson and N.G.A. Ralph (eds), *Palaeobiological investigations. Research designs, methods and data analysis*. Oxford, British Archaeological Reports, 105–19

Jones, A.K.G. 1990. 'Experiments with fish bones and otoliths: implications for the reconstruction of past diet and economy' in D.E. Robinson (ed.), *Experimentation and reconstruction in environmental archaeology*. Oxford, Oxbow Books, 143–6

Jones, G.E.M. 1992. 'Weed phytosociology and crop husbandry: identifying a contrast between ancient and modern practice', *Review of Palaeobotany and Palynology* 73, 133–43

Jones, K.T. and Metcalfe, D. 1988. 'Bare bones archaeology: bone marrow indices and efficiency', *Journal of Archaeological Science* 15, 415–23

Jones, M. 1986. *England before Domesday*. London, Batsford

Keegan, W.F. 1989. 'Stable isotope analysis of prehistoric diet' in M.Y. Iscan and K.A.R. Kennedy (eds), *Reconstruction of life from the skeleton*. New York, Alan Liss, 223–36

Keller, F. 1878. *The lake dwellings of Switzerland and other parts of Europe*. London, Longmans

Kennard, A.S. 1943. 'The post-Pliocene non-marine mollusca of Hertfordshire', *Transactions of the Hertfordshire Natural History Society and Field Club* 22, 1–18

Kent, S. 1992. 'The current forager controversy: real versus ideal views of hunter-gatherers', *Man* 27, 45–70

Kenward, H.K. 1978. *The analysis of archaeological insect assemblages: a new approach*. London, Council for British Archaeology

Kerney, M.P. 1977. 'A proposed zonation scheme for Late-Glacial and Post-Glacial deposits using land mollusca', *Journal of Archaeological Science* 4, 387–90

Kerney, M.P., Preece, R.C., and Turner, C. 1980. 'Molluscan and plant biostratigraphy of some Late Devensian and Flandrian deposits in Kent', *Philosophical Transactions of the Royal Society* B 291, 1–43

Kirk, W. 1989. 'Historical geography and the concept of the behavioural environment' in F.W. Boal and D.N. Livingstone (eds), *The behavioural environment: essays in reflection, application and re-evaluation*. London, Routledge, 18–30

Kopec, R.J. 1963. 'An alternative method for the construction of Thiessen polygons', *Professional Geographer* 15, 24–6

Kozarski, S., Gonera, P., and Antczak, B. 1988. 'Valley floor development and paleohydrological changes: the Late Vistulan and Holocene history of the Warta River (Poland)' in G. Lang and C. Schluchter (eds), *Lake, mire and river environments*. Rotterdam, Balkema, 185–203

Krebs, C.J. 1985. *Ecology: the experimental analysis of distribution and abundance*. New York, Harper & Row

Küster, H. 1991. 'Phytosociology and archaeobotany' in D.R. Harris and K.D. Thomas (eds), *Modelling ecological change*. London, Institute of Archaeology, 17–26

Lack, D. 1954. *The natural regulation of animal numbers*. Oxford, Clarendon Press

Lageras, P. and Sandgren, P. 1994. 'The use of mineral magnetic analyses in identifying middle and late Holocene agriculture: a study of peat profiles in Smaland, southern Sweden', *Journal of Archaeological Science* 21, 687–97

Lalueza Fox, C. and Perez-Perez, A. 1994. 'Dietary information through the examination of plant phytoliths on the enamel surface of human dentition', *Journal of Archaeological Science* 21, 29–34

Lam, P.K.S. and Calow, P. 1988. 'Differences in the shell shape of *Lymnaea peregra* (Müller) (Gastropoda: Pulmonata) from lotic and lentic habitats; environmental or genetic variance?' *Journal of Molluscan Studies* 54, 197–207

Langran, G. 1993. *Time in geographic information systems*. London and Washington DC, Taylor & Francis Inc.

Laville, H., Rigaud, J., and Sackett, J. 1980. *Rock shelters of the Perigord: geological stratigraphy and archaeological succession*. New York, Academic Press

Lee, R.B. and DeVore, I. (eds) 1968. *Man the hunter*. New York, Aldine

Leroi-Gourhan, A. 1968. 'Le Neanderthalien IV de Shanidar', *Bulletin de la Société prehistorique Français* 15, 79–83

Lewin, J., Davies, B.E., and Wolfenden, P.J. 1977. 'Interactions between channel change and historic mining sediments' in K.J. Gregory (ed.), *River channel changes*. Chichester, John Wiley, 353–67

Lewin, J., Macklin, M.G., and Woodward, J.C. (eds) 1995. *Mediterranean Quaternary river environments*. Rotterdam, Balkema

Lilley, J.M., Stroud, G., Brothwell, D.R., and Williamson, M.H. 1994. *The Jewish burial ground at Jewbury*. York, Council for British Archaeology

Limbrey, S. 1975. *Soil science and archaeology*. London, Academic Press

Limbrey, S. 1992. 'Micromorphological studies of buried soils and alluvial deposits in a Wiltshire river valley' in S. Needham and M.G. Macklin (eds), *Alluvial archaeology in Britain*. Oxford, Oxbow Books, 53–64

Limbrey, S. and Robinson, M. 1988. 'Dryland to wetland: soil resources in the upper Thames valley' in P. Murphy and C. French (eds), *The exploitation of wetlands*. Oxford, British Archaeological Reports, 129–41

Lobb, S. 1995. 'Excavations at Crofton causewayed enclosure', *Wiltshire Archaeological and Natural History Magazine* 88, 18–25

Lock, G. and Stancic, Z. (eds) 1995. *Archaeology and GIS: a European perspective*, London, Taylor & Francis

Löffler, H. 1986. 'Ostracod analysis' in B.E. Berglund (ed.), *Handbook of Holocene palaeoecology and palaeohydrology*. Chichester, John Wiley, 693–702

Lowe, J.J. and Walker, M.J.C. 1997. *Reconstructing Quaternary environments*, 2nd edn. Harlow, Longman

Lucas, A.T. 1960. *Furze: a survey and history of its uses in Ireland*. Dublin, The Stationery Office

Lyman, R.L. 1994. *Vertebrate taphonomy*. Cambridge, Cambridge University Press

Lynn, C. 1983. 'Some "early" ring-forts and crannogs', *Journal of Irish Archaeology* 1, 47–58

Macan, T.T. 1963. *Freshwater ecology*. London, Longman

McBryde, I. 1986. 'Artefacts, language and social interaction: a case study from south-eastern Australia' in G. Bailey and S. Callow (eds), *Stone-age prehistory: studies in memory of Charles McBurney*. Cambridge, Cambridge University Press, 77–93

McCarthy, M.R. 1995. 'Archaeological and environmental evidence for the Roman impact on vegetation near Carlisle, Cumbria', *The Holocene* 5(4), 491–5

McDowell, P.F. 1983. 'Evidence of stream response to Holocene climatic change in a small Wisconsin watershed', *Quaternary Research* 19, 100–16

Macklin, M.G. 1985. 'Floodplain sedimentation in the upper Axe valley, Mendip, England', *Transactions of the Institute of British Geographers* 10, 235–44

Macphail, R.I. 1991. 'The archaeological soils and sediments' in N.M. Sharples, *Maiden Castle excavations and field survey 1985–6*. London, English Heritage, 106–18

Macphail, R.I. 1992. 'Soil micromorphological evidence of past soil erosion' in M.G. Bell and J. Boardman (eds), *Past and present soil erosion: archaeological and geographical perspectives*. Oxford, Oxbow Books, 197–215

Macphail, R.I. 1994. 'The reworking of urban stratigraphy by human and natural processes' in A.R. Hall and H.K. Kenward (eds), *Urban-rural connexions: perspectives from environmental archaeology*. Oxford, Oxbow Books, 13–43

Macphail, R.I. and Goldberg, P. 1995. 'Recent advances in micromorphological interpretations of soils and

REFERENCES

sediments from archaeological sites' in A.J. Barham and R.I. Macphail (eds), *Archaeological sediments and soils*. London, University College, 1–24e

Macphail, R.I., Courty, M.-A., and Gebhardt, A. 1990. 'Soil micromorphological evidence of early agriculture in north-west Europe', *World Archaeology* 22, 53–69

Magurran, A.E. 1988. *Ecological diversity and its measurement*. London, Croom Helm

Mannion, A. 1978. 'Late Quaternary deposits from Southeast Scotland II: the diatom assemblage of a marl core', *Journal of Biogeography* 5, 301–18

Marshall, F. 1994. 'Food sharing and body part representation in Okiek faunal assemblages', *Journal of Archaeological Science* 21, 65–77

Marshall, F. and Pilgram, T. 1991. 'Meat versus within-bone nutrients: another look at the meaning of body-part representation in archaeological sites', *Journal of Archaeological Science* 18, 149–63

Maschener, H.D.G. (ed.) 1996. *New methods, old problems: GIS in modern archaeological research*. Urbana, Southern Illinois University

Matheson, C. 1939. 'A survey of the status of *Rattus rattus* and its subspecies in the seaports of Great Britain and Ireland', *Journal of Animal Ecology* 8, 84

Matiskainen, H. and Alhonen, P. 1984. 'Diatoms as indicators of provenance in Finnish sub-Neolithic pottery', *Journal of Archaeological Science* 11, 147–58

Matthews, W. 1995. 'Micromorphological characterisation and interpretation of occupation deposits and microstratigraphic sequences at Abu Salabikh, Southern Iraq' in A.J. Barham and R.I. Macphail (eds), *Archaeological sediments and soils*. London, University College, 41–74

Matthews, W., French, C.A.I., Lawrence, T., Cutler, D.F., and Jones, M.K. 1997. 'Microstratigraphic traces of site formation processes and human activities', *World Archaeology*, 29, 281–308

Mayr, E. 1970. *Populations, species and evolution*. Cambridge, Mass., Belknap Press

Mercer, R. and Tipping, R. 1994. 'The prehistory of soil erosion in the northern and eastern Cheviot Hills, Anglo-Scottish borders' in S. Foster and T.C. Smout (eds), *The history of soils and field systems*. Aberdeen, Scottish Cultural Press, 1–25

Mielke, H.W. 1989. *Patterns of life*. London, Unwin Hyman

Mills, C.M., Crone, A., Edwards, K.J., and Whittington, G. 1994. 'The excavation and environmental investigation of a sub-peat stone bank near Loch Portain, North Uist, Outer Hebrides', *Proceedings of the Society of Antiquaries of Scotland* 124, 155–71

Mitchell, F.J.G. 1988. 'The vegetational history of the Killarney oakwoods, s-w Ireland: evidence from fine spatial resolution pollen analysis', *Journal of Ecology* 76, 415–36

Mitchell, F.J.G. 1990. 'The impact of grazing and human disturbance on the dynamics of woodland in s-w Ireland', *Journal of Vegetational Science* 1, 245–54

Molloy, K. and O'Connell, M. 1993. 'Early land use and vegetation history at Derryinver Hill, Renvyle Peninsula, Co. Galway, Ireland' in F.M. Chambers (ed.), *Climate change and human impact on the landscape*. London, Chapman and Hall, 186–99

Moore, P.D. 1993. 'The origin of blanket mire, revisited' in F.M. Chambers (ed.), *Climate change and human impact on the landscape*. London, Chapman and Hall, 218–44

Moore, P.D., Webb, J.A., and Collinson, D. 1991. *An illustrated guide to pollen analysis*, 2nd edn. London, Hodder & Stoughton

Morales Muñiz, A. and Senz Bretón, J.L. 1994. 'Arqueo-acarología: potencialides y limitaciones de una prácticamente inédita subdisciplina arqueózoologica', *Pyrenae* 25, 17–29

Morgan, A.V. 1987. 'Late Wisconsin and early Holocene palaeoenvironments of east-central North America based on assemblages of Coleoptera' in W.F. Ruddiman and H.E. Wright jr (eds), *North America and adjacent oceans during the last glaciation*. Boulder, Colorado

Morrison, I. 1985. *Landscape with lake dwellings: the crannogs of Scotland*. Edinburgh, Edinburgh University Press

Munsell soil colour chart (revised edn.) 1994. Baltimore, Maryland: Munsell Color Company Inc.

Mytum, H.C. 1988. 'On-site and off-site evidence for changes in subsistence economy: Iron Age and Romano-British west Wales' in J.L. Bintliff, D.A. Davidson and E.G. Grant (eds), *Conceptual issues in environmental archaeology*. Edinburgh, Edinburgh University Press, 72–81

REFERENCES

Needham, S.P. 1991. *Excavation and salvage at Runnymede Bridge, 1978: the Late Bronze Age and waterfront site*. London, British Museum Press

Needham, S. and Macklin, M.G. (eds) 1992. *Alluvial archaeology in Britain*. Oxford, Oxbow Books

Needham, S. and Spence, T. 1997. 'Refuse and the formation of middens', *Antiquity* 71, 77–90

Newnham, R.M. 1992. 'A 30,000-year pollen, vegetation and climate record from Otakairangi (Hikurangi), Northland, New Zealand', *Journal of Biogeography* 19, 541–54

Newton, S. and Porter, D. 1988. *Modernisation frustrated: the politics of industrial decline in Britain since 1900*. London, Unwin Hyman

Nicholson, A.J. 1933. 'The balance of animal populations', *Journal of Animal Ecology* 2, 131–78

Nicholson, A.J. 1954. 'An outline of the dynamics of animal populations', *Australian Journal of Zoology* 2, 9–65

Nicholson, R.A. 1996. 'Bone degradation, burial medium and species representation: debunking the myths, an experiment-based approach', *Journal of Archaeological Science* 23, 513–33

O'Connor, T.P. 1993. 'Process and terminology in mammal carcass reduction', *International Journal of Osteoarchaeology* 3, 63–7

O'Connor, T.P. 1997. 'Working at relationships: another look at animal domestication', *Antiquity* 71, 149–56

Odum, E.P. 1971. *Fundamentals of ecology*, 3rd edn. Philadelphia, W.B. Saunders Co.

Olsen, S.L. and Shipman, P. 1988. 'Surface modification on bone: trampling *versus* butchery', *Journal of Archaeological Science* 15, 535–53

Ommanney, F.D. 1938. *South Latitude*. London, Longman Green & Co

Orme, B. 1981. *Anthropology for archaeologists*. London, Duckworth

Orton, C. 1980. *Mathematics in archaeology*. London, Collins

Payne, S. and Munson, P.J. 1985. 'Ruby and how many squirrels? The destruction of bones by dogs' in N.R.J. Fieller, D.D. Gilbertson and N.G.A. Ralph (eds), *Palaeobiological investigations: research design, methods and data analysis*. Oxford, British Archaeological Reports, 31–40

Pearsall, D.M. 1988. 'Interpreting the meaning of macroremain abundance: the impact of source and context' in C.A. Hastorf and V.S. Popper (eds), *Current palaeoethnobotany*. Chicago, University of Chicago Press, 97–118

Pearsall, D.M. 1989. 'Adaptation of prehistoric hunter-gatherers to the high Andes: the changing role of food resources' in D.R. Harris and G.C. Hillman (eds), *Foraging and farming*. London, Unwin Hyman, 318–32

Pearsall, D.M. and Piperno, D. (eds) 1993. *Current research in phytolith analysis: applications in archaeology and palaeoecology*. Philadelphia, MASCA

Peglar, S.M. 1993. 'The mid-Holocene *Ulmus* decline at Diss Mere, Norfolk, UK: a year-by-year pollen stratigraphy from annual laminations', *The Holocene* 3 (1), 1–13

Perry, D., Buckland, P.C., and Snaesdottir, M. 1985. 'The application of numerical techniques to insect assemblages from the site of Storaborg, Iceland', *Journal of Archaeological Science* 12, 335–45

Pielou, E.C. 1975. *Ecological diversity*. New York, John Wiley

Pielou, E.C. 1984. *The interpretation of ecological data*. New York, John Wiley

Piperno, D.R. 1988. *Phytolith analysis: an archaeological and geological perspective*. San Diego, Academic Press

Pollard, E., Hooper, M.D., and Moore, N.W. 1974. *Hedges*. London, Collins

Pope, K.O. and van Andel, T.H. 1984. 'Late Quaternary alluviation and soil formation in the southern Argolid: its history, causes and archaeological implications', *Journal of Archaeological Science* 11, 281–306

Powers-Jones, D.R. 1994. 'The use of phytolith analysis in the interpretation of archaeological deposits: an Outer Hebridean example' in R. Luff and P. Rowley-Conwy (eds), *Whither environmental archaeology?* Oxford, Oxbow Books, 41–50

Price, P.W. 1975. *Insect ecology*. New York, John Wiley

Putman, R.J. 1994. *Community Ecology*. London, Chapman and Hall

Rackham, O. 1986. *The history of the countryside*. London, J.M. Dent & Sons Ltd

REFERENCES

Rackham, O. 1990. *Trees and woodland in the British landscape*, revised edn. London, J. M. Dent & Sons Ltd

Rapp, G. and Mulholland, S.C. (eds) 1992. *Phytolith systematics: emerging issues*. New York, Plenum Press

Reeves-Smyth, T. and Hamond, F. 1983. *Landscape archaeology in Ireland*. Oxford, British Archaeological Reports

Reitz, E.J., Newsom, L.A., and Scudder, S.J. 1996. *Case studies in environmental archaeology*. New York, Plenum Press

Renfrew, C., Gimbutas, M., and Elster, E. (eds) 1986. *Excavations at Sitagroi: a prehistoric village in northeast Greece, I*. Los Angeles, University of California

Reynolds, P. 1994. 'Butser ancient farm' in A.P. Fitzpatrick and E.L. Morris (eds), *The Iron Age in Wessex: recent work*. Salisbury, Wessex Archaeology, 11–14

Rigaud, J.-P., Simek, J.F., and Ge, T. 1995. 'Mousterian fires from Grotte XVI (Dordogne, France)', *Antiquity* 69, 902–12

Roberts, N. 1989. *The Holocene, an environmental history*. Oxford, Blackwell

Robinson, D. (ed.) 1990. *Experimentation and reconstruction in environmental archaeology*. Oxford, Oxbow Books

Robinson, M. 1988. 'Molluscan evidence for pasture and meadowland on the floodplain of the upper Thames basin' in P. Murphy and C. French (eds), *The exploitation of wetlands*. Oxford, British Archaeological Reports, 101–12

Robinson, M. and Lambrick, G.H. 1984. 'Holocene alluviation and hydrology in the upper Thames basin', *Nature* 308, 809–14

Roig jr, F., Roig, C., Rabassa, J., and Boninsegna, J. 1996. 'Fuegian floating tree-ring chronology from subfossil *Nothofagus* wood', *The Holocene* 6(4), 469–76

Rose, J., Boardman, J., Kemp, R.A. and Whiteman, C.A. 1985. 'Palaeosols and the interpretation of the British Quaternary stratigraphy' in K.S. Richards, R.R. Arnett and S. Ellis (eds), *Geomorphology and Soils*. London, George Allen and Unwin, 348–75

Rosen, A.M. 1986. *Cities of clay: the geoarchaeology of tells*. Chicago, University of Chicago Press

Rosen, A.M. 1994. 'Identifying ancient irrigation: a new method using opaline phytoliths from emmer wheat', *Journal of Archaeological Science* 21, 125–32

Rossignol, J.A. and Wandsnider, L. (eds) 1992. *Space, time and archaeological landscapes*. New York, Plenum Press

Rouse, A. and Evans, J.G. 1994. 'Modern land Mollusca from Maiden Castle, Dorset, and their relevance to the interpretation of subfossil archaeological assemblages', *Journal of Molluscan Studies* 60, 315–29

Rust, A. 1943. *Die alt- und mittelsteinzeitliche Funde von Stellmoor*. Neumunster, Karl Wachholtz Verlag

Sadler, J.P. and Jones, J.C. 1997. 'Chironomids as indicators of Holocene environmental change in the British Isles' in A.C. Ashworth, P.C. Buckland and J.P. Sadler (eds), *An inordinate fondness for beetles. Studies in Quaternary entomology*. Chichester, John Wiley, 219–32

Sandor, J.A. 1992. 'Long-term effects of prehistoric agriculture on soils: examples from New Mexico and Peru' in V.T. Holliday (ed.), *Soils in archaeology: landscape evolution and human occupation*. Washington, Smithsonian Institution Press, 217–45

Saunders, P. 1985. 'Space, the city and urban sociology' in D. Gregory and J. Urry (eds), *Social relations and spatial structures*. London, Macmillan, 67–89

Schelvis, J. 1987. 'Some aspects of research on mites (Acari) in archaeological samples', *Palaeohistoria* 29, 211–18

Schelvis, J. 1992. 'The identification of archaeological dung deposits on the basis of remains of predatory mites (Acari; Gamasida)', *Journal of Archaeological Science* 19, 677–82

Schirmer, W. 1988. 'Holocene valley development on the upper Rhine and Main' in G. Lang and C. Schluchter (eds), *Lake, mire and river environments*. Rotterdam, Balkema, 153–60

Schofield, A.J. 1991. 'Lithic distributions in the upper Meon valley: behavioural response and human adaptation on the Hampshire chalklands', *Proceedings of the Prehistoric Society* 57 (2), 159–78

Sczeicz, J. and MacDonald, G.M. 1996. 'A 930-year ring-width chronology from moisture sensitive white spruce (*Picea glauca* Moench) in northwestern Canada', *The Holocene* 6 (2), 345–51

REFERENCES

Sernander, R. 1908. 'On the evidence of post-glacial change of climate furnished by peat mosses of northern Europe', *Geologisk Foreningens Stockholm Forhandlingar* 30, 465–73

Shackley, M.L. 1975. *Archaeological sediments: a survey of analytical methods*. London, Butterworth

Shanks, M. and Hodder, I. 1995. 'Processual, postprocessual and interpretive archaeology' in I. Hodder et al. (eds), *Interpreting archaeology: finding meaning in the past*. London, Routledge, 3–29

Sharples, N.M. 1991. *Maiden Castle: excavations and field survey 1985–6*. London, English Heritage

Shennan, S. 1988. *Quantifying archaeology*. Edinburgh, Edinburgh University Press

Shipman, P. 1981. *Life history of a fossil: an introduction to vertebrate taphonomy and palaeoecology*. Cambridge, Mass., Harvard University Press

Simmons, I.G. 1997. *Humanity and environment. A cultural ecology*. Harlow, Addison Wesley Longman Ltd

Simmons, I.G. and Cundill, P. 1974. 'Late Quaternary vegetational history of the North Yorkshire Moors I: pollen analyses of blanket peats', *Journal of Biogeography* 1, 159–69

Simmons, I.G., Atherden, M.A., Cloutman, E.W., Cundill, P.R., Innes, J.B., and Jones, R.L. 1993. 'Prehistoric environments' in D.A. Spratt (ed.), *Prehistoric and Roman archaeology of north-east Yorkshire*. London, Council for British Archaeology, 15–50

Simmons, I.G., Turner, J., and Innes, J.B. 1990. 'An application of fine-resolution pollen analysis to later Mesolithic peats of an English upland' in C. Bonsall (ed.), *The Mesolithic in Europe*. Edinburgh, Edinburgh University Press, 206–17

Singer, R. and Wymer, J.J. 1982. *The Middle Stone Age at Klasies River Mouth in South Africa*. Chicago, Chicago University Press

Singh, G. and Smith, A.G. 1973. 'Post-glacial vegetational history and relative land- and sea-level changes in Lecale, Co. Down', *Proceedings of the Royal Irish Academy* B 73, 1–51

Smith, A.G. 1965. 'Problems of inertia and threshold related to Post-Glacial habitat changes', *Proceedings of the Royal Society* B 161, 331–42

Smith, A.G. and Cloutman, E.W. 1988. 'Reconstruction of Holocene vegetation history in three dimensions at Waun Fignen Felen, an upland site in South Wales', *Philosophical Transactions of the Royal Society* B 322, 159–219

Smith, H. 1996. 'An investigation of site formation processes on a traditional Hebridean farmstead using environmental and geoarchaeological techniques' in D. Gilbertson, M. Kent and J. Grattan (eds), *The Outer Hebrides: the last 14,000 years*. Sheffield, Sheffield Academic Press, 195–206

Smith, R. 1984. 'The ecology of Neolithic farming systems as exemplified by the Avebury region of Wiltshire', *Proceedings of the Prehistoric Society* 50, 99–120

Smith, R.J.C., Healey, F., Allen, M.J., Morris, E.L., Barnes, I., and Woodward, P.J. 1997. *Excavations along the route of the Dorchester bypass, Dorset, 1986–8*. Salisbury, Wessex Archaeology

Soil Survey Staff 1975. *Soil taxonomy, a basic system of soil classification for making and interpreting soil surveys*. Washington DC, United States Department of Agriculture

Soil Survey Staff 1990. *Keys to soil taxonomy*, 4th edn. Bladesburg, Virginia, Pocahontas Press

Southwood, T.R.E. 1978. *Ecological methods*. London, Chapman and Hall

Sparks, B.W. 1986. *Geomorphology*, 3rd edn. Harlow, Longman

Spoerry, P. 1992. *Geoprospection in the archaeological landscape*. Oxford, Oxbow Books

Spratt, D.A. and Simmons, I.G. 1976. 'Prehistoric activity and environment on the North York Moors', *Journal of Archaeological Science* 3, 193–210

Stafford jr, T.W. and Semken, H.A. 1990. 'Accelerator 14C dating of two micromammal species representative of the Late Pleistocene disharmonious fauna from Peccary Cave, Newton County, Arkansas', *Current Research in the Pleistocene* 7, 129–32

Stahl, A.B. 1995. 'Has ethnoarchaeology come of age?', *Antiquity* 69, 404–7

Stahl, P.W. 1996. 'The recovery and interpretation of microvertebrate bone assemblages from archaeological contexts', *Journal of Archaeological Method and Theory* 3 (1), 31–75

Stallibrass, S. 1990. 'Canid damage to animal bones: two current lines of research' in D.E. Robinson (ed.), *Experimentation and reconstruction in environmental archaeology*. Oxford, Oxbow Books, 151–65

Stevenson, J.B. 1975. 'Survival and discovery' in J.G. Evans, S. Limbrey and H. Cleere (eds), *The effect of man on the landscape: the highland zone*. London, Council for British Archaeology, 104–8

REFERENCES

Steward, J. 1955. *A theory of culture change*. Urbana, University of Illinois Press

Stewart, K.M. 1991. 'Modern fishbone assemblages at Lake Turkana, Kenya: a methodology to aid in recognition of hominid fish utilisation', *Journal of Archaeological Science* 18, 579–603

Stone, G.D. 1993. 'Agricultural abandonment: a comparative study in historical ecology' in C.M. Cameron and S.A. Tomka (eds), *Abandonment of settlements and regions: ethnoarchaeological and archaeological approaches*. Cambridge, Cambridge University Press, 74–81

Stukeley, W. 1743. *Abury, a temple of the British Druids, with some others, described*. London

Sukopp, H., Blume, H.-P., and Kunick, W. 1979. 'The soil, flora and vegetation of Berlin's waste lands' in I.C. Laurie (ed.), *Nature in cities*. Chichester, John Wiley, 115–32

Sutcliffe, A.J. 1985. *On the track of Ice Age mammals*. London, British Museum (Natural History)

Tankard, A.J. and Schweitzer, F.R. 1976. 'Textural analysis of cave sediments: Die Kelders, Cape Province, South Africa' in D.A. Davidson and M.L. Shackley (eds), *Geoarchaeology: earth science and the past*. London, Duckworth, 289–316

Tauber, H. 1965. 'Differential pollen dispersion and the interpretation of pollen diagrams', *Danmarks Geologiske Undersøgelse* IIR 89, 1–69

Tauber, H. 1967. 'Investigations of the mode of pollen transfer in forested areas', *Revue of Palaeobotany and Palynology* 3, 277–87

Taylor, J.J., Innes, J.B., and Jones, M.D.H. 1994. 'Locating prehistoric wetland sites by an integrated palaeoenvironmental/geophysical survey strategy at Little Hawes Water, Lancashire' in R. Luff and P. Rowley-Conwy (eds), *Whither environmental archaeology?* Oxford, Oxbow Books, 13–23

Thieme, H. 1997. 'Lower Palaeolithic hunting spears from Germany', *Nature* 385, 807–10

Thomas, J. 1990. 'Silent running: the ills of environmental archaeology', *Scottish Archaeological Review* 7, 2–7

Thomas, J. and Whittle, A. 1986. 'Anatomy of a tomb: West Kennet revisited', *Oxford Journal of Archaeology* 5, 129–56

Thomas, K.D. 1982. 'Neolithic enclosures and woodland habitats on the South Downs in Sussex, England' in M. Bell and S. Limbrey (eds), *Archaeological aspects of woodland ecology*. Oxford, British Archaeological Reports, 147–70

Thomas, K.D. 1985. 'Land snail analysis in archaeology: theory and practice' in N.R.J. Fieller, D.D. Gilbertson and N.G.A. Ralph (eds), *Palaeobiological investigations: research design, methods and data analysis*. Oxford, British Archaeological Reports, 131–56

Thomas, K.D. 1989. 'Hierarchical approaches to the evolution of complex agricultural systems' in A. Milles, D. Williams and N. Gardner (eds), *The beginnings of agriculture*. Oxford, British Archaeological Reports, 55–76

Thompson, R. and Oldfield, F. 1986. *Environmental magnetism*. London, Allen & Unwin

Tipping, R. 1992. 'The determination of cause in the generation of major prehistoric valley fills in the Cheviot Hills, Anglo-Scottish border' in S. Needham and M.G. Macklin (eds), *Alluvial archaeology in Britain*. Oxford, Oxbow Books, 111–21

Tixier, H., Duncan, P., Scehovic, J., Yani, A., Gleizes, M., and Lila, M. 1997. 'Food selection by European roe deer (*Capreolus capreolus*): effects of plant chemistry, and consequences for the nutritional value of their diets', *Journal of Zoology*, London 242 (2), 229–46

Toll, M.S. 1988. 'Flotation sampling: problems and some solutions, with examples from the American Southwest' in C.A. Hastorf and V.S. Popper (eds), *Current palaeoethnobotany*. Chicago, University of Chicago Press, 36–52

Tomka, S.A. 1993. 'Site abandonment behaviour among transhumant agro-pastoralists: the effects of delayed curation on assemblage composition' in C.M. Cameron and S.A. Tomka (eds), *Abandonment of settlements and regions: ethnoarchaeological and archaeological approaches*. Cambridge, Cambridge University Press, 11–24

Trigger, B.G. 1989. *A history of archaeological thought*. Cambridge, Cambridge University Press

Trudgill, S. 1989. 'Soil types. A field identification guide', *Field Studies* 7, 337–63

Turner, J. 1975. 'The evidence for land use by prehistoric farming communities: the use of three-dimensional pollen diagrams' in J.G. Evans, S. Limbrey and H. Cleere (eds), *The effect of man on the landscape: the highland zone*. London, Council for British Archaeology, 86–95

REFERENCES

Tyldesley, J.B. 1973. 'Long range transmission of tree pollen to Shetland. I. Sampling and trajectories', *New Phytologist* 72, 175–81

van Zeist, W. and Bottema, S. 1977. 'Palynological investigations in western Iran', *Palaeohistoria* 19, 19–85

van Zeist, W., van Hoorn, T.C., Bottema, S., and Woldring, H. 1976. 'An agricultural experiment in the unprotected saltmarsh', *Palaeohistoria* 18, 111–53

Vincent, J.F.V. 1982. *Structural biomaterials*. London, Macmillan Press

Vita-Finzi, C. and Higgs, E.S. 1970. 'Prehistoric economy in the Mount Carmel area of Palestine: catchment analysis', *Proceedings of the Prehistoric Society* 36, 1–37

von Grafenstein, U., Erlenkeuser, H., Muller, J., and Kleinmann-Eisenmann, A. 1992. 'Oxygen isotope records of benthic ostracods in Bavarian lake sediments', *Naturwissenschaften* 79, 145–52

von Post, L. 1916. 'Om skogstradspollen i sydsvenska torfmosselagerfoljder (foredragsreferat)', *Geologisk Forening, Stockholm* 46 (Published in translation as Davis, M.B. and Faegri, K. 1967. 'Forest tree pollen in south Swedish peat bog deposits', *Pollen et Spores* 9, 375–401.)

Wainwright, G.J. 1972. 'The excavation of a late Neolithic enclosure at Marden, Wiltshire', *Antiquaries Journal* 52, 177–239

Walker, D. 1970. 'Direction and rate in some British Post-Glacial hydroseres' in D. Walker and R.G. West (eds), *Studies in the vegetational history of the British Isles*. Cambridge, Cambridge University Press, 117–39

Walker, M.J.C., Griffiths, H.I., Ringwood, V., and Evans, J.G. 1993. 'An early Holocene pollen, mollusc and ostracod sequence from the lake marl at Llangorse lake, South Wales, U.K.', *The Holocene* 3 (2), 138–49

Walker, S.T. 1880. 'Report on the shell heaps of Tampa Bay, Florida', *Annual Report of the Smithsonian Institution*, 1879, 413–22

Walker, S.T. 1885. 'Mounds and shell-heaps on the west coast of Florida', *Annual Report of the Smithsonian Institution*, 1883, 854–68

Wedel, W. 1953. 'Some aspects of human ecology in the Central Plains', *American Anthropologist* 55 (4), 499–514

West, R.G. 1972. *Pleistocene geology and biology*. London, Longman

White, R. 1983. 'Changing land-use patterns around the Middle/Upper Palaeolithic transition: the complex case of the Perigord' in E. Trinkaus (ed.), *The Mousterian legacy*. Oxford, British Archaeological Reports, 113–21

Whittington, G. 1983. 'A palynological investigation of a second millennium BC bank system in the Black Moss of Achnacree', *Journal of Archaeological Science* 10, 283–91

Whittle, A.W.R. 1993. 'The Neolithic of the Avebury area: sequence, environment, settlement and monuments', *Oxford Journal of Archaeology* 12, 29–53

Whittle, A.W.R. 1994. 'Excavations at Millbarrow Neolithic chambered tomb, Winterbourne Monkton, north Wiltshire', *Wiltshire Archaeological and Natural History Magazine* 87, 1–53

Whittle, A.W.R., Rouse, A.J., and Evans, J.G. 1993. 'A Neolithic downland monument in its environment: excavations at Easton Down long barrow, Bishops Cannings, North Wiltshire', *Proceedings of the Prehistoric Society* 59, 197–240

Wilde, W. 1862. 'Upon the unmanufactured animal remains belonging to the Academy', *Proceedings of the Royal Irish Academy* 7, 181–211

Wilkinson, T.J. 1988. 'The archaeological components of agricultural soils in the Middle East: the effects of manuring in antiquity' in W. Groenman-van Waateringe and M. Robinson (eds), *Man-made soils*. Oxford, British Archaeological Reports, 93–114

Wilson, D.R. 1975. 'The evidence of air-photographs' in J.G. Evans, S. Limbrey and H. Cleere (eds), *The effect of man on the landscape: the Highland Zone*. London, Council for British Archaeology, 108–11

Wilson, D.R. 1982. *Air photo interpretation for archaeologists*. London, Batsford

Wolda, H. 1981. 'Similarity indices, sample size and diversity', *Oecologia* 50, 296–302

Woodward, P.J. 1991. *The south Dorset ridgeway: survey and excavations 1977–84*. Dorchester, Dorset Natural History & Archaeological Society

REFERENCES

Wyman, J. 1868. 'An account of the fresh-water shellheaps of the St Johns River, East Florida', *American Naturalist* 2, 393–403, 440–63

Wymer, J.J. 1962. 'Excavations at the Maglemosian sites at Thatcham, Berkshire, England', *Proceedings of the Prehistoric Society* 28, 329–61

Wymer, J.J. 1976. 'The interpretation of Palaeolithic cultural and faunal material found in Pleistocene sediments' in D.A. Davidson and M.L. Shackley (eds), *Geoarchaeology: earth science and the past*. London, Duckworth, 327–34

Wymer, J.J. 1992. 'Palaeoliths in alluvium' in S. Needham and M.G. Macklin (eds), *Alluvial archaeology in Britain*. Oxford, Oxbow Books, 229–34

Yalden, D.W. 1983. 'Yellow-necked mice in archaeological contexts', *Bulletin of the Peakland Archaeological Society* 33, 24–9

Yalden, D.W. 1995. 'Small mammals from Viking-age Repton', *Journal of Zoology, London* 237, 655–57

Yalden, D.W. and Morris, P.A. 1990. *The analysis of owl pellets*. London, The Mammal Society

Zangger, E. 1992. 'Neolithic to present soil erosion in Greece' in M.G. Bell and J. Boardman (eds), *Past and present soil erosion: archaeological and geographical perspectives*. Oxford, Oxbow Books, 133–47

Zeder, M.E. 1991. *Feeding cities*. Washington, Smithsonian Institution Press

Zeuner, F.E. 1952. 'Palaeobotanical aspects of geochronology', *The Palaeobotanist* 1, 448–55

Zeuner, F.E. 1963. *A history of domesticated animals*. London, Hutchinson

INDEX

abandonment of sites and ecosystems 76, 79, 152–3, 185, 187–8
abundance 23–6
Abu Salabikh, Iraq 153
acid oxic environments 80
actualistic data 104–5, 115, 132
aerial photography 109, 118
affordances 216
age-morphologic groups (AMGs) 126–7
aggregation,
    of fauna 21–5, 46
    of mineral particles 121
agro-pastoralists 184–5, 191
algal cysts 136
alkaline oxic environments, *see* basic oxic environment
allochthonous assemblages 73, 76, 87
alluviation 44, 77, 170, 208, 212–13
alluvium 11, 88–9, 130, 160–1, 164
altitudinal gradients 48–50
analogy 9, 170, 181–2
    continuous 182, 185
    direct historical 185
    formal 170, 181
    relational 170, 181
Anglo-Saxon estates 110–12
anoxic environments 81, 134, 137, 146, 151, 199
Apodemus flavicollis 133
archaeological and environmental theory 3–8
archaeological features and sites 149–54
archipelagos 95
argillic brownearths 123–4
arid and semi-arid environments 33, 43, 50, 114, 120, 138, 154, 161–2, 209–10
art 191
Ascaris lumbricoides 145
assemblage comparison 175–7
Asvan Kale, Turkey 153
atmosphere 10, 12–13
Aufnahmen 178–9

Australia 133, 183
autochthonous assemblages 73, 76, 87
autogenic sediments 40, 89
Avebury, Wiltshire, area and henge 88, 202–4, 209
azonal soils 35

backswamp deposits 160–1
basic oxic environments 80–1, 141, 146
basins 118, 131, 157
basin size,
    in lake and sediment formation 38
    in pollen sampling 70–1
    in sampling 102, 135
bedding in sediments 121
beetles 140–1, 177–9
Beverley, Yorkshire 103–4
biochemical studies 17–18
bioenergetics 28, 45
biogenic calcification 42, 122, 161 (*see also* travertine; tufa)
biogenic sediments 40–3, 89, 149
biogeography 14, 18, 116–17, 133, 148
biological activity 31–45, 82 (*see also* biogenic sediments; bioturbation)
biological indicators 40, 60–1, 80–1, 83, 132–47
biomass 26, 28–9, 45
biomolecules 40, 148
biosphere 10, 13–15, 17–29
bioturbation 83, 86–8, 125
bird-of-prey pellets 69
birds and bird bones 23–5, 72–3, 146–7
blank areas 90–1, 109, 115, 117
blanket bog 155–9 (*see also* peat and peat bogs)
Boeotia survey 100, 194
bogs, *see* blanket bog; peat and peat bogs
Bolivia 191
bone and bones 69, 74–5, 86, 145–8, 185, 187

bovids 133
Boxgrove, Sussex 166
bran 136
Bronze Age archaeology 157–8, 210–11
Bronze Age erosion 209
brownearth soils 31, 33–5, 205, 208–9
building debris and buildings 113–14, 152–3
bulk laboratory analysis of soils and sediments 121–2
burial 57, 67, 185
buried soils 87–8, 119, 149–51, 153, 156–7, 160–2, 165, 202–7
burning of vegetation, *see* charcoal; fire
Burren, Co. Clare 97
butchery 74, 86
Butser experimental farm, Hampshire 188

cairns 156–7, 210–11
calcareous sediments 134, 149, 161, 166 (*see also* sand dunes and blown sand; travertine; tufa)
calcification and calcium carbonate 32–3, 121–2, 141 (*see also* biogenic calcification)
Canada 133, 139
Cantabria 190–1
carbon isotopes 148 (*see also* radio-carbon dating)
carbon:nitrogen (C:N) ratios 31, 121
Carlisle 196–7, 199–200
carnivores 74–5
Castillo, Spain 190–1
catenas 51, 91
causation and causes 130–1, 133, 169, 180, 212–14
causewayed camps 207
cave deposits and caves 86, 96, 138–9, 190–1
    river 164–5
    sea 167
cellulose 137

INDEX

censuses 110
central place theory 194
cereals 136–7, 186
chalklands 90, 101, 162, 201–9
change 41, 47, 54–6, 82–9
charcoal 124–5, 137–9, 152, 156–8
    (see also fire)
charters 110, 115
Chile 141
chironomids 145
chitin 140
chronology,
    of archaeological and
        environmental studies 2–3
    of prehistoric deposits 4–5
chronospecies 17
circumstantial evidence 130, 170, 213–14
Cladocera 145
clastic sediments and clasts 39–42, 84
clay 124
clay-humus complexes 31–2
climate and climate change 12, 14, 33, 37, 43, 64, 133, 141, 143, 149, 179
cluster analysis 176
coasts 49–50, 91, 98, 155, 165–8, 190–1
Coleoptera, see beetles
collagen 145–6
colluviation 44, 87
colluvium 35, 88–9, 160–2, 199
colour of soils and sediments 120
communities 26–9, 178
companion animals 183
competitive exclusion 133
complexity of communities 28–9, 45
computer simulation 194–5
contexts 102, 148–68
conversation 183
coral 12, 40,149
corers and cores for analysis 99, 120, 142, 156
correlations,
    of data 171–2, 180
    of sediment histories and archaeology 129–31, 180
    of soil distributions and archaeology 126–9, 210–11
    (see also distributions)
coversands 48, 166
crannogs 4, 60–6, 79, 93, 102
crumb structure of soil 29, 31, 130, 208–9
cryoturbation 43, 83, 87, 119
cultivation terraces 209–10 (see also lynchets)

cultural ecology 6
cultural relativism 15–16
Cumbria 158
curation 114–15, 187
cut-marks on bones 75

Dalmore, Lewis, Outer Hebrides 168
Danebury, southern Britain 154
dark earth 152
Dassenetch 183, 185, 187
data analysis 171–80
data and inferences 104–5
dating 197–8 (see also dendrochronology; radiocarbon dating)
daub 152
death 57, 67–77
    attritional 68
    catastrophic 69
    community 72
    ecosystem 76–7
    effects of predators and scavengers on 69
    hominid- and human-originated 69, 72
    locationally biased 68–9, 85
death assemblages 66–77
deathtraps 68–9
decalcification of chalk soils 204, 209
Deeping St Nicholas, Lincolnshire 123
demes 18–19
dendrochronology 62, 139, 161, 199
deposition 39, 78–92, 187 (see also sedimentation; structured deposition)
Des Moines valley, USA 126–7
diagenic 40, 57
diatomite 136
diatoms 136
diet 148
discard 114–15, 187
disequilibrium communities and ecosystems 28, 35, 54, 179
dispersal 18, 21–5, 46
distal factors (in interpretation) 169, 204
distributions,
    archaeological 106–17, 127–8, 187
    biological 17–28, 116–17, 133–4
    disjunct 19, 133
    of bones 187
    of resources 46, 108
    of sedimentation, sediments and soils 43,128
    relationships between soils and

archaeology 110–12, 128–9, 210–11
    (see also spatial associations; spatial distribution)
ditches 83–4, 96, 150–1, 154, 187–8, 205
Dithmarschen, Schleswig–Holstein 165
diversity 172–5
    inventory 28, 46, 175, 203–4
    of artefacts on land surfaces 89
    of ecosystems 47–51
    of land 174–5
    of sediments 11, 175
diversity indexes 141, 172–4
DNA 17–18, 40, 148
documents 84, 95, 109–12, 182
dogs 183
Domesday Book 110
droppings, see faeces
dry valleys, see valleys
dung, see faeces
dusty clay coatings 124–5

earthworks 90–1, 150, 153–4, 156
    (see also monument areas and monuments)
earthworms 86, 88, 117, 152
East Brittany survey 100, 111
Easton Down, Wiltshire 206–9
ecological groupings 177–8, 203–5
ecology 1–3, 8, 17–44, 75
economic prehistory 5
ecophenotypes 20–1, 29, 149
ecosystems 1, 7–8, 45–6, 76
    change 46–7
    classification 54
    death 76
ecotones 53, 165
eggs,
    of nematodes 145
    of snakes 104
eluviation 32–3
electronic distance measurers (EDMs) 113
elm decline 133
elytra 140
enchytraeid worms 152
energy pyramids 20
environments,
    global 10–16
    of burial 80–2
    of crannogs 60–5
    operational 15–16
    perceived 15–16, 184–5, 201, 215–16
    reconstruction of 62–3, 140, 147
    sampling of 101–2

ephippia 145
equifinality 170
equitability 28–9, 172–5
erosion 13, 36, 39–41, 50, 89, 114, 130–1, 158–9, 161, 167, 208–10, 212–14
essences 216
estuaries 165–7
ethnoarchaeology 122, 185–9
ethnography, see human-life studies
ethnohistory 182
European Pollen Monitoring Programme 71
eurytopic species 18–19
evolution 2–4, 148
experiment 75, 122, 151, 181–2, 185–9
experimental crop production 188, 210

faeces 136
    human 145, 200–1
fans, alluvial, gravel and outwash 160–1
farms 186
feedback 169, 171
field boundaries 156–7, 159, 206 (see also earthworks; lynchets; walls)
fieldwalking 114–15
fire 124, 137–8, 170, 183
fishbones 75
Fisher's alpha diversity index (α) 174–5
floodloams 122 (see also alluvium)
floodplains 73, 77, 84, 151, 160–4
flooring and floors 151–4
food-sharing 74–5
food webs 29, 45
form-fossils 133
founder effect 19
French caves 164

game theory 194
Geest 48, 165–6
genetic diversity 18–19, 23
geographical information systems (GIS) 106–7, 192–4
geology,
    drift 11, 106–8, 118
    solid 11, 106–8, 154
geophysics 115–16
glaciation 2, 4, 48
gleying and gleys 31–2, 35, 120
gradation and gradients 12, 43, 47–51
gradiometer survey 116

grasslands 76, 155, 192–3, 204, 207
Greece 117, 211 (see also Boeotia survey)
'guilt by association' 170 (see also circumstantial evidence)

habitats 18–19, 45, 177–8
Hadrian's Wall 196–8
haymeadows 178
Hekla, Iceland 139
hedgerow dating 116–17
hillforts 153–5, 210–11
history 184
Holocene ecosystem change 55
    modelling 192–3
    sequences 165–6, 205–9
holotypes 132
honey 200
horses 201
Howmore, South Uist 186
human activities 79, 94–5
human-life studies 75, 182–5, 191
humans 1, 13, 15–16, 36–7, 53, 67–9, 72, 133–4, 145–6, 182–6, 216–18
humus, humus formation 31–2, 121, 152
hunter-gatherers 183–4, 191
Hvar, Croatia 210–11
hydroseres 52–3
hydrosphere 10

Iceland 139, 141
identification of biological materials 132–3 (see also chronospecies)
illuviation 124–5
increments 148–9, 153, 161
inductivism 170
inferences 104–5
infra-red imaging 109, 122–3
Inner Hebrides 166
inorganic raw materials 46, 108, 148–9
insects 61–4, 151 (see also beetles)
integration of data, methods and scales,
    in field surveys 117
    of molluscs and ostracods 143
    of plant macrofossils 138–9
    of pottery and valley sediments 89, 161
    of snails and small mammals 146 (see also spatial associations; spatial distribution)
internal grouping of data 172–7
interpretation 131, 147, 169–70, 181
    of beetles 140–1

of different spatial scales 64–6, 196, 209 (see also spatial associations; spatial distribution)
    of mixed plant assemblages 138–9
    of vertebrates 146–7
interpretational distances 104–5
interstitial fauna 40, 63, 86 (see also bioturbation)
intertidal deposits 165–6
intra-site scales 141
inventory diversity, see diversity
invertebrates 140–5
Ireland 71, 97, 155
Iron Age archaeology 153–5, 192, 197–8, 201, 210–11
islands 155 (see also archipelagoes)
isostasty 162, 165–6
isotopes 40, 102, 148–9

Jos plateau, Nigeria 184

Karst 35, 97, 155, 160, 210
Kennard, Alfred 5, 141
Klasies River Mouth, South Africa 167
K-strategists 20–1, 23
Kubiena boxes 99, 122

lakes and lake deposits 45, 48, 52–3, 60–6, 71, 79, 129, 136, 142–5, 156–9, 165–6
Lake Turkana, Kenya 187
Lake Zeribar, Iran 135
landblocks 80–2, 95, 100
landform sediment assemblages (LSAs) 127
landscape archaeology 7, 215
land snails, see molluscs
land-use 90
    in relation to archaeology 106, 109–11, 128–9, 209–14
    maps 108
languages and linguistics 148, 185 (see also toponymy)
Late-glacial deposits and sequences 119, 165–6
laterisation 32–3
layering in vegetation 45
leaching 32–5, 50
Lejre experimental farm, Denmark 188
lessivation 32–3, 35, 125
life 9
limestone areas 164–5
lipids 148
litho-pedoturbation 87–8 (see also bioturbation)

239

INDEX

lithosphere 10–12
lithostratigraphical units 41
Little Hawes Water, Lancashire 135
Little Ice Age 13
localities 60–5
local scale of study 64, 114, 135, 137, 158, 187–9, 191, 209–10
loess 48
loss on ignition 121
lynchets 35–6, 99, 130–1, 157, 159–60 (*see also* cultivation terraces)

machair 166
magnetic susceptibility 115, 122 (*see also* mineral magnetics)
Maiden Castle, Dorset 154–5, 178, 206, 209
Malham, Yorkshire 35, 51, 192–3
manuring 136, 151, 186, 212
maps 106–10, 118, 169
marginal land 56, 76, 102, 155, 167
Marsch 166
Marsworth, Hertfordshire 41
mathematical methods 171
mathematical modelling 194–5
maximisation 182–3
Mediterranean environments 49, 114, 120, 210–12
Mesolithic archaeology 134, 156–8, 166–7, 205, 207
metal analyses 121, 148–9
metalworking areas 154
methods 93, 104–5
Mexico 191
micromorphology of soils 99, 120, 122–5, 131, 152, 214
middens 151–2, 166, 186–7 (*see also* shell middens)
mineralisation 32, 208
mineral magnetics 121, 157
mires, *see* peat and peat bogs
mites 144
mobility 21–5, 64, 67–9, 141 (*see also* movement)
 of people 183–4
modelling,
 of bone deposition 74–5
 of death assemblages 70–5
 of environmental change 107
 of Holocene vegetation 192–3
 of pollen deposition 70–1
molecules 40, 148–9
molluscs,
 land and freshwater 5, 18, 20–2, 63–4, 83, 86, 102, 132, 141–4,

146–7, 151–3, 166–7, 173, 177, 201–9
 marine 21–2, 166–7
monument areas and monuments 95, 99–101, 201–2, 205
Moor 166
Morisita-Horn index 175–6
moss polsters 71
movement 82–9
 of economic materials 89, 149, 199–200, 212
 (*see also* mobility)
mudbricks 152–3
Munsell Colour System 120
mutual climatic range method (MCRM) 179–80

Namib 146
nearest neighbour analysis 172, 194
Nehrungen 166
nematodes 145
Neolithic archaeology 201–9
neutral oxic environments 81
New Mexico 209–10
New Zealand 133
niches 23, 27, 45, 177–8, 182, 215–18
North America 4–7, 127, 141, 146, 183–4
northern England 196–201
North York Moors 135, 152
Nothofagus 133
numerical analyses 141, 144

occupation layers 151–2
offsite archaeology and areas 16, 60, 94, 102–3, 113, 117, 153–4
Olduvai Gorge 42–3
open fields 110
ordination techniques 176–7
Oribatidae 144
Oryctolagus cuniculus 133
ostracods 63–4, 142–4, 171
Outer Hebrides 166–8, 186
ova, *see* eggs
oxic environments 80–2, 137
oxidation, oxidation state 32, 127
oxygen 80–2
oxygen isotopes 149

palaeochannels, *see* river channels
Palaeolithic archaeology,
 Lower 164, 167, 170
 Upper 124, 142, 190–2
panmictic units 18
particle-size analysis 121
patterned ground 43

peat and peat bogs 52, 70–1, 97, 139, 155–8, 165–7, 197–8, 213
 for fuel 200
Pech-de-l'Azé, France 124
pedozones 124–5
people, *see* humans
periglacial deposits and features 43, 48, 83, 85, 119, 150, 161–2, 203 (*see also* cryoturbation)
Perm 209–10
pH 80–2, 121
phosphate measurements 115
physico-chemical conditions 80–2
phytoliths 135–6, 186
phytosociology 28, 178–9
pits 69, 84–6, 151–5
Pitstone soil 119, 127
placenames, *see* toponymy
plaggen soils 37
plagioseres 53
plant groupings 13–14 (*see also* vegetation)
plant macrofossils 4, 137–9, 151
plaster 152–3
ploughmarks 198, 203, 208
podsolisation 32–3, 35
polarisation of debates 130, 170, 213
pollen and pollen analysis 4, 64, 70–1, 99, 102–3, 120, 129, 133–5, 197–8
 in three dimensions 135, 157
 in uplands 157–8
 of anthropogenic sediments 135
 of soils 135
 of urban sites 135, 199–200
polymorphism 19
polyphenism 21
populations 18–26
 abundance 23–4
 distribution 17–18, 21–3
possibilistic methods 105, 115, 181–95
post-deposition 57–8, 78–92
pottery,
 history 58
 in field-walking 114
 provenancing 136, 149
precipitates and precipitation 33, 39–40, 42–3, 89, 149
predators 69
prediction of distributions 107, 126–8
pre-peat soils and walls 95–7, 156
preservation 78, 80–2
 conditions 113
 of sites 80, 91–2, 95–6, 187
procedures of research 93

INDEX

productivity 28, 45, 182
proximal factors (in interpretation) 169
proxy data 104–5, 132
pulse stability 53

quarry hollows 153–5

radar surveys 109
radiocarbon dating (C-14) 122, 129, 140, 146, 161, 198
raised beaches 166–7
Rancholabrea, California, tar seeps 69
rank-order curves 172–4
recording systems 106
reduction 32
refugia xiv
regional scale of study 60–5, 107, 135, 157, 189–91, 201–2, 210–14
relationships between life and death 76–7
representativeness,
    of air photograph data 109
    of death assemblages 76–7
    of preservation 92
research areas, priorities and strategies 6–7, 93–4, 105, 117, 167, 196
resolution,
    of archaeology 78
    of chronologies 196–8
    of preservation 91, 101, 114, 117
risk management 183–4
river channels 160–1, 164
river deposits 84, 109
River Kennet 164, 202, 205
river terraces 51, 126–8, 158–64
River Thames 164
river valleys, *see* valleys
Romans 110–11, 192–3
    impact of in northern England 196–201
rooms 141, 153
r-strategists 20–1, 23
rubbish 58, 86, 151, 187
Runnymede Bridge, Surrey 164, 178

salinisation 32–3
saltmarsh plants 200–1
sampling 7, 87, 98, 101–3, 120, 150, 152
    judgemental 107, 118
    strategies 93–105, 107
sand dunes and blown sand 84, 165–8 (*see also* coversands)

sandurs 48
satellite photographs 109, 118
scavengers and scavenging 69, 74–5, 187
Schöningen, Germany 170
sea-level changes 165–7, 194
seasonality 149
sedimentation 37–44, 89
sediments 11, 13, 38–44, 62, 80, 83, 88
    distinction from soils 88, 118–19
    histories and sequences 43, 129–31, 175
    mapping and analysis 118–31
seeds 137–9, 148
segmentation (of environment) 34, 46
semi arid environments, *see* arid and semi-arid environments
seres 53
sewers 199–200
Shannon–Wiener index (H') 172–4, 203
shell chemistry 141, 149
shell middens 4, 149, 166
shells, *see* molluscs
shell shape 20–2
Shetlands 1, 70, 98–9, 122
similarity indices 175–6
simplification of data 171
Simpson's lambda index (λ) 174
Sitagroi, Greece 153
site catchments and SCA 66, 102–3, 189
site exploitation territorial analysis (SETA) 5, 189–94, 210–11
sites and monuments records (SMRs) 108–9
sites and site scale of study 7, 16, 60–6, 98–9, 187–91, 194, 209–10
small vertebrates 146
snails, *see* molluscs
soil,
    and archaeology 127–31
    and experimental archaeology 188
    and land-use 209–14
    analysis 118–31, 210
    classification 13, 29–37, 80, 83, 86, 88
    description 120–1
    distinction from sediments 88, 119
    formation 29–34
    histories 129–31, 209–14
    horizons 30–2, 36–7, 120
    maps 35, 37, 106–8, 118–21, 192
    processes 29–37

profiles 32, 36
sampling 99
    (*see also* buried soils)
solifluxion deposits 38, 42, 120
sols lessivés 35
southern African caves 164–5
southern England 201–9
South Nesting, Shetland 98–9
South Street, Wiltshire 150, 202, 205–9
South Uist 185–6
space 46, 60–2
Spain 190–1
spatial associations,
    of archaeology, demography and alluviation 212–13
    of sites and territories 189–94
    of sites and vegetation 192–3, 210–11
    (*see also* temporo-spatial associations)
spatial dislocations 199–201
spatial distribution 110, 172–80, 189–94 (*see also* distributions)
spatial integrity and resolution 65, 135, 140
spatial scales,
    of interpretation 142–3, 196, 209–14
    of research 60–2, 94–100 (*see also* regional scale of study; sites and site scale of study)
    of sampling 100–2
spatial variation,
    in biological materials and middens 143–4, 152, 206
    in plant fossils 138–9
    in tells 152–3
    of oxygen isotope signals 149
    of preservation 187–9
    of vertebrate remains 146–7
special nature of sites 65–6
species, species concept 17, 132–3
species' ecology 17–26, 177–80, 204–5
species' ranges 17–23, 133, 177, 179–80 (*see also* distributions, biological)
sponge spicules 199
sporopollenin 134
statistics 171–2
Stellmoor, Germany 117, 142–3
stenotopic species 19
stone artefacts,
    chemical analysis of 148–9
    in field walking 114
storm beaches 165–6

241

INDEX

stress,
  in *Daphnia* 145
  in trees 139
strontium:calcium ratios 148
structured deposition 85–6, 153, 184
submerged forests 165–7
succession 47, 51–3, 179
sumps, *see* basins
survey,
  above ground 112–13
  ground-level 112–16
  in relation to archaeological preservation 113
swamps 73
synanthropic species 133–4, 151

taphonomy 7, 57–9, 188–9
  in relation to diversity indexes 174
  of artefacts 114–15
  of charcoal 137–8
  of plant macrofossils 138–9
  of soil and sediments 127–8, 150–1, 204
  of urban sites 199–201
  (*see also* spatial associations; spatial dislocations)
tectonic change 11, 13, 165
Tehuacán Valley, Mexico 138
tells 152–3
temperature 81–2, 149
temporo-spatial scales,
  association of agriculture, settlement and alluviation 129–31, 212–13
  dislocation of settlement and land-use 212–13
  of structure and variation 24–5, 72–3, 187, 196
terrace agriculture 209–10
territories and catchments 66, 183, 185, 189–94

texture,
  of grain surfaces 121
  of soils and sediments 120–1
therapy 183
Thiessen polygons 194, 211
tidal increments 149
till 38, 42, 48, 120
timber, *see* wood
time 9, 46–7
  in ecosystems 46–7, 52, 101, 165
  mapping 110–12
  mapping by GIS 106–7
  in sediments 40
  in soils 124, 126–7
  sequences in tells 153
  short-time events 188–9
Tierra del Fuego, tree-ring chronology 139
tooth-marks on bones 75
topogenous mires 157–8 (*see also* peat and peat bogs)
topography 13, 106, 131
toponymy 109–12
towns and cities, *see* urban archaeology and sites
tradition 184
trampling 75, 187
travertine 41
trees 139
tree stumps in peat 139
tree-throw pits 87–8, 124–5, 150, 203–5
trophic levels 26–9
tsunami deposits 1, 136
tufa 42, 124
tunnel valleys 48, 142
Trichuris trichiura 145

underwater surveys 109
uplands 49–51, 90–1, 154–9, 209–10

urban archaeology and sites 86, 100, 135–6, 152, 178, 199–201

valleys,
  dry 161–2, 207
  long profiles of 164
  river 51, 90, 95–6, 126–8, 130–1, 160–5, 205–7
vegetation 14, 178–9, 192–3
vertebrates 145–7 (*see also* bones)
visibility of archaeology and preservation 78, 90, 98, 1678, 198–9, 201
volcanoes 65, 139

walls 152–3, 156, 197
Warta River, Poland 161
water conditions 145
waterlogging, *see* anoxic environments
weather 12–13, 139, 212
weathering 29–31, 39, 120, 151, 187–8
weeds 138–9, 178
wells 151
West Penwith, Cornwall, 157
Wierde 166
Wilsford Shaft, Wiltshire 151
Wiltshire 201–9
wood 139–40, 149, 161, 163, 197, 199, 212
woodlands and woodland clearance 72–3, 117, 129, 133–5, 139–40, 146, 158, 197–9, 203–7, 212

York 196–7, 199–201

Zea mays 136, 210
zones of destruction and preservation 91–2